THE BIRTH OF ETHICS

From the time of conception, through the gestation of pregnancy, to the birth of a newborn child exists an extraordinary, emergent ethics. How does this ethics come into being when a child is conceived? How does the appearance of ethics in pregnancy differ from its emergence after birth? How does the original meaning of ethics relate to modern morality in decision making?

In this book, Michael van Manen explores these ethical moral complexities and conceptualizations of life's beginnings. He delves into perennial and contemporary aspects of conception, pregnancy, and birth to present ethics as a fundamental phenomenon in the experiential encounter between parent and child. Even in the context of neonatal-perinatal medicine, where all manner of medical technologies and illnesses may potentially complicate the developing relation of parent and child, ethics is always already present yet also enigmatic in its origin. And yet, to approach ethical moral questions, we need to understand the inception of ethics.

The Birth of Ethics: Phenomenological Reflections on Life's Beginnings is an essential text not only for health professionals and researchers but also for parents, family members, and others who care and take responsibility for newborns in need of medical care.

Michael van Manen is an Associate Professor in the Department of Paediatrics, Faculty of Medicine & Dentistry, and Endowed Chair of Health Ethics and Director of the John Dossetor Health Ethics Centre at the University of Alberta, Canada. He has a clinical practice as a physician in neonatal-perinatal medicine with the Stollery Children's Hospital.

PHENOMENOLOGY OF PRACTICE

Series Editor: Max van Manen
University of Alberta

The series *Phenomenology of Practice* sponsors books that are steeped in phenomenological scholarship and relevant to professional practitioners in fields such as education, nursing, medicine, pedagogy, clinical and counseling psychology. Texts in this series distinguish themselves for offering inceptual and meaningful insights into lived experiences of professional practices, or into the quotidian concerns of everyday living. Texts may reflectively explicate and focus on aspects of method and dimensions of the philosophic and human science underpinnings of phenomenological research.

For further manuscript details available from the Series Editor: please contact Max van Manen at vanmanen@ualberta.ca / +250-294 4345

Other volumes in this series include:

Visual Phenomenology
Encountering the Sublime Through Images
Erika Goble

Pedagogical Tact
Knowing What to Do When You Don't Know What to Do
Max van Manen

Classic Writings for a Phenomenology of Practice
Michael van Manen and Max van Manen

Phenomenology of the Newborn
Life from Womb to World
Michael van Manen

For a full list of titles in this series, please visit **www.routledge.com**

THE BIRTH OF ETHICS

Phenomenological Reflections on Life's Beginnings

Michael van Manen

Routledge
Taylor & Francis Group

NEW YORK AND LONDON

First published 2021
by Routledge
52 Vanderbilt Avenue, New York, NY 10017

and by Routledge
2 Park Square, Milton Park, Abingdon, Oxon OX14 4RN

Routledge is an imprint of the Taylor & Francis Group, an informa business

© 2021 Michael van Manen

Library of Congress Cataloging-in-Publication Data
A catalog record for this title has been requested

ISBN: 978-0-367-56065-2 (hbk)
ISBN: 978-0-367-62742-3 (pbk)
ISBN: 978-1-003-11059-0 (ebk)

Typeset in Bembo
by Taylor & Francis Books

For Miep

CONTENTS

Acknowledgments   *ix*

Introduction   1

Section I
Before a Child is Born   13

Conceiving Ethics   15

Pregnant with Child   21

Ultrasound Imaging and Virtual Presence   27

Nascent Expectations and Hope   33

Meaningful Outcomes   39

Section II
In the Cradle of the Newborn Intensive Care Unit   45

Newborn Encounters   47

Technics of Touch   53

Skin-to-Skin Togetherness   59

Imaging the Child   63

Attachment and Responsibility   67

Section III
Ethics and Decisions

Section III
Ethics and Decisions   71

Situating Decisions   73

Decisions Without Choices   77

Looking for Ways Out of Decisions   82

Thinking and Feeling Through Decisions   87

Decisions and Indecisions   91

Falling into Decisions   96

Conclusion and Beginnings   *99*
Ethical Moral Perspectives   *108*
References   *117*
Index   *127*

ACKNOWLEDGMENTS

I thank the parents and families who participated in the research studies informing this book, and for their willingness to share their often very personal stories
and life experiences with me. The anecdotes contained in this book are drawn
from their stories. While the anecdotal accounts have been fictionalized to preserve the anonymity of the children, parents, and staff members, I very much
hope that this text speaks authentically both to their joys and difficulties of parenting a child who has spent time in the neonatal intensive care unit of the
hospital.

I thank the practicing trainees, bedside nurses, respiratory therapists, dieticians,
pharmacists, social workers, nurse practitioners, and physicians in Neonatal-Perinatal Medicine at the Stollery Children's Hospital for their support. And, I am very
appreciative of my friends and colleagues at the John Dossetor Health Ethics
Centre. In particular, I thank Wendy Austin, Catherine Adams, Paul Byrne,
Pamela Brett-MacLean, Po-Yin Cheung, Chloe Joynt, Brendan Leier, Gillian
Lemermeyer, Ernest Phillipos, and Juzer Tyebkhan, whose comments and conversations helped me develop this book. I also thank the Routledge editor in chief
Hannah Shakespeare for her support and encouragement.

I am grateful for the funding received from the Canadian Institutes for Health
Research and the Women & Children's Health Research Institute that made this
work possible. I thank my family, including my mother and father, and my brother
Mark, who have supported me since before I could utter words. And I thank the
love of my life Miep, and our incredible children Luka and Jude, not only for their
patience when I work long hours but also for always reminding me of what is most
important when I come home at the end of each day.

This book aims to unfold the ethics of life's beginnings. I hope it stimulates
the reader's wondering, thoughtfulness, and sensitivity, and the affirmation of
health care professionals concerned with the ethics of clinical practice in

neonatal-perinatal medicine, but also and especially those parents and their precious newborns who are faced with such challenging life beginnings.

Section 1 includes material based on an article published as:

Van Manen, M. A. (2019). "Uniqueness and novelty in phenomenological inquiry." *Qualitative Inquiry, 26(5)*, 486–490.

Section 2 includes material based on articles published as:

Van Manen, M. A. (2012). "Ethical responsivity and pediatric parental pedagogy." *Phenomenology & Practice, 6(1)*, 5–17.
   Van Manen, M. A. (2012). "Technics of touch in the neonatal intensive care." *Medical Humanities, 38(2)*, 91–96.
   Van Manen, M. A. (2015). "The ethics of an ordinary medical technology." *Qualitative Health Research, 25(7)*, 996–1004.

Section 3 includes material based on an article published as:

Van Manen, M. A. (2014). "On ethical (in)decisions experienced by parents of infants in neonatal intensive care." *Qualitative Health Research, 24(2)*, 279–287.

INTRODUCTION

In neonatal-perinatal medicine, our tiniest of patients start their newborn lives with us in complicated circumstances. Rarely does anyone plan to have a baby born premature. Families generally do not anticipate that their pregnancy is fated to be interrupted, that they potentially will be confronted by grave decisions for a child who they have not even had the chance to hold. And yet, the ethics that we need to speak of—those ethics that we can speak of being born—would seem to need our considered attention.

When she developed contractions at 24 weeks gestation, the sensation was completely foreign to her. It was her first pregnancy. She did not know what it was, perhaps something she ate, a bladder infection, or something else? The idea that it might be labor did not even enter her mind. It was only when her water broke in the earliest hours of morning that she realized she had to get to the hospital. On arriving, there was no time for conversations with specialists about 'what to do.' Doctors, nurses, and others filled the room. Cody was born. He appeared small and scrawny, weighing slightly less than 600 grams. They, surrounded him, in a flurry of orchestrated medical actions, and placed him on a radiant warmer. She could not see what was going on as they initiated resuscitation, supporting his breathing with a face mask. She could not see them inserting a breathing tube to assist with the inflation of his immature lungs. She could not see them insert long intravenous lines into his belly button for administration of fluids and medications. She could only tell that they were working on him as the medical team sequestered him from sight. And then as quick as the team arrived, they were gone. They had taken him away with hurried, placating words: "We are bringing him to the newborn intensive care unit (the NICU). He seems to be doing ok." A couple hours later, Cody's mother had recovered enough to come to the neonatal intensive care unit to see him. He had been placed in an incubator. The nurse encouraged her to lay a hand on his body, which was adorned with probes, wires, and intravenous lines. His skin was sticky, shiny, and glistening, almost to the point of being translucent. His hands and tiny fingers were delicately flexed into small fists, quivering ever so slightly in response to her touch. His mouth was spread, fixed open by the breathing tube. His eyes were not visible, still fused-closed from immaturity; she sensed he felt her presence. His smooth facial features wrinkled in response to her touch. She could not find words to respond to his expression, but she recognized his need for soothing. And then the doctor came, sitting down with her at the bedside. After some talk, the conversation came to a choice: to continue on with medical treatments or to stop. The doctor said it was her decision, a choice they would have liked to have discussed in advance of the medical interventions already done—given the burden of intensive care, and the risks of complications or death. Such conversations had not been possible given Cody's precipitous delivery. She stared at Cody. This is not what she would have ever wanted for her child. But in seeing him and touching him she 'knew' that she did not have a choice.

This is a story of ethics at the beginning of life. It is an account of Cody's birth, a child born near the threshold of viability, at a moment when parents are normally expected to make decisions for their child. But the mother of Cody is charged with a different kind of decision that does not feel like it is a decision for her. And yet, it is a decision that needs an answer for the doctor. The doctor hopes that such ambivalent decisions are made for the child based on what the parents perceive to be in their child's 'best interests.' The medical staff hopes that the 'right' decision is the one that places the child's wellbeing first and foremost (Buchanan & Brock, 1989). But how does one determine the right decision? Is this not little more than a surface understanding of ethics and morality? Is this truly the starting point for ethics? The truth is that before the moral question of 'what to do?' is even posed, ethics has already announced itself. The physician's moral question is asked of an already existent ethics—an ethics founded in the mother's primal experience of being 'addressed' or 'felt' by her child.

In the neonatal intensive care unit, when this new mother is encouraged to touch her newborn, she senses the vulnerability of her child. The needful child does not need communicate what it needs for his mother to experience its needfulness. As she sees her baby, and as she touches the delicate skin, the mother is touched in return. She senses the need for soothing. This is a profoundly ethical moment. Even in neonatal-perinatal medicine, where all manner of medical technologies and illnesses potentially complicate the relation of parent and child, where life itself may come into the world prematurely, we find ethics expressed in terms of need and response: responsibility—this vulnerable newborn's life makes a claim on the mother through the reciprocal sensation of touch and being touched.

Ethics is an enigmatic phenomenon: How did this ethics come into being? Was it born in this moment of the newborn touch? Or is this ethics always already there? What signs of ethics exist at conception and persist through pregnancy? This child, who is new to this world of neonatal intensive care, may already have been expected or anticipated. He or she may already have been given a name. One wonders, what is the origin of this life, newly born from the womb? Did it come from nonexistence, nihility, nothingness? And yet, we recognize that an inceptual presence of ethics has already made itself known in this moment of touch.

So, when the doctor arrives, asking for a decision, we realize that the possibility of choice does not really exist in this ethical moment. This kind of ethics is not commensurable with deliberating questions of morality: 'What is to be done?' Should we be surprised if the mother cannot entertain anything other than wanting her child to live? When seeing and touching him, she has been touched by an otherness that transcends her subjectivity.

> Something did not seem quite right. Mohamed, born at only 26 weeks gestation and now 2 weeks old, was having frequent pauses in his breathing. The nurse and doctor assessed, deciding to try administering a load of caffeine to stimulate his breathing. Mohamed's condition worsened—deep and frequent drops in his heart rate and saturations were now occurring. His parents were called. On exam, his abdomen appeared blue, loopy, and distended. X-ray imaging showed findings consistent with extensive diseased bowel—necrotizing enterocolitis. Mohamed's feeds were held and a breathing tube was put in. The pace of care quickened: antibiotics ordered, vasoactive medication infusions started, and surgery consulted. When Mohamed's parents arrived to the nursery from home, they could tell immediately that the situation was grave. His skin was pale and mottled. His breathing was unnaturally rhythmic. He seemed to make no response to their presence. An exploratory laparotomy was recommended. The parents consented. Mohamed was taken for surgery. A long hour of waiting, and then he was back from the operating room appearing no better than before. "There is nothing I can do," said the surgeon. "Too much of bowel has been affected. Virtually all is necrotic. He cannot survive." The neonatal team spoke with the parents, counseling them to shift their goals of care to focus exclusively on Mohamed's comfort—to discontinue medical interventions. The parents declined. They were not ready. They believed he had a chance. Reluctantly, the neonatal team agreed to continue, expecting that Mohamed would die in the coming hours on the breathing machine. Oxygen needs increased. His heart rate lost variability. Blood pressure dropped. Kidney failure ensued. The medical team was sure that Mohamed was dying. Again, they discussed with the parents removing the breathing tube. The parents were steadfast in their resolve to carry on. So, the medical

team continued, adjusting his sedative and analgesic medications, to try to ensure some comfort, expecting the end to be near. But, over the course of days, Mohamed did not die. His color improved. He appeared to breathe on his own between the breaths afforded by the breathing machine. Still, the medical team again counseled the parents to discontinue medical interventions. They said that even if Mohamed's heart, lungs, and other vital organs recovered, if he continued to live, that he would need intravenous nutrition for months to years. They explained that there existed the very possibility there had been significant injury to his brain. That although the condition of his heart, lungs, and kidneys seemed better, the overall situation had actually not improved. It had only changed. Mohamed could not be fed and was dependent on intravenous nutrition. However, the doctors conceded there was an uncertain and remote possibility that Mohamed could survive to grow, that one day he might be big enough for bowel transplantation. His parents wanted to carry on. They prayed that some greater spiritual entity would intervene. While the team was not encouraged by the family's spiritual appeal, they accepted it was possible, reasonable, to continue intensive care. Mohamed had a chance, however remote it might be. And yet, the team was conflicted and struggling. For them, Mohamed was suffering through these medical interventions and the possibility of a good outcome was just so very unlikely.

What are the ethics of neonatal-perinatal medicine when a child's bodily being is affected by illness? How does the technomedical reality of intensive care shape ethics? How do healthcare professionals experience the presence of the ethical?

Ethics is commonly regarded as a branch of philosophy. But philosophy itself is a cultural phenomenon—a *theoria*—that started with the early Greek. Ethics is more originary as a human phenomenon than philosophy. And, certainly, as an existential phenomenon, ethics is more primordial than what drives the contemporary forms of bio-ethics. Etymologically, the term 'ethics' originates from *ethos*, meaning "the moral atmosphere of a people or a community" (Klein, 1971, p. 260). The Online Etymology Dictionary (OED) gives the etymological meaning of ethos as "the 'genius' of a people, characteristic spirit of a time and place," indicating that ethics and ethos are phenomena deeply embedded in the ancient history of humanity.

The arcane meaning of this word 'ethos,' and certainly the Greek use of this word, is not, however, merely limited to a subjective sense of a person's habit, temperament, or disposition. Rather, *ethos* transcends virtue, conveying a sense of receptivity. For example, the power of music to influence its hearer's emotions, behaviors and even morals for the Greeks was an *ethos* whereby the *ethos* of music is neither wholly in the melody that moves the listener nor solely in the listener who is stirred by song (Weiss & Taruskin, 1994). Instead, it is the intertwining of the given and the response that appears as *ethos*. Heidegger finds the meaning of ethos in the sense of a dwelling place: "the open region in which the human being dwells" (1998/1946, p. 269). It is a clearing for the possibility of a revealing of being with what is other to ourselves.

In neonatal-perinatal medicine, the intertwining of the given and response, identity and alterity, self and other, adult and child, dwell in a technomedical reality (van Manen, 2012a). For healthcare professionals, their encounter with a child is composed by technomedical 'intentionality' such that they see, hear, and are touched by a child through both the most mundane and most complicated of medical interventions and clinical measures. While the power of the technomedical apparatus enables knowledge that explains, predicts, and offers the potential to direct treatment, it also fundamentally shapes the ethics of neonatal-perinatal medicine. The nurse takes action when noticing that the infant's heart rate slows and saturations fall. The doctor suspects pain when diagnosing diseased or compromised organs. The team may see suffering in a child who is dependent on invasive medical technologies. The NICU is implicated in a technomedical environment, where ethics will make its appearance. And, of course, the technomedical apparatus also affects visitors of the patient, as the healthcare team guides the family to see just what the practitioners see.

Some families of sick children rely on spirituality to provide hope. But from the rational point of view of science, it may sometimes look like a hopeless hope. Whether the spiritual is connected or not to a divine faith, this too may look like an appearance of ethics. The spiritual offers promise of purpose and meaning. And yet, if we want to understand the primal significance of ethics, then we need to consider that ethics should neither be reduced to science nor faith, neither to rationality nor to feeling or sentimentality.

A newborn infant is born in an unexpected very poor condition. No breathing effort can be detected. No pulse can be felt. There are no signs of life. The delivery room nurses start mask ventilation and chest compressions. A 'Code Pink' is announced overhead, calling for help. The neonatal team arrives within minutes. The baby is intubated, and chest compressions are continued, umbilical lines inserted, and resuscitation medications administered. There is no response: The baby remains apneic and pulseless. More medications are given, and breathing support and chest compressions continue. The baby continues to appear lifeless. The parents are pale, their faces wet with sweat and tears. The doctor and nurses know this baby will not survive, that attempts at resuscitation ought to stop. The doctor looks to the parents, as she prepares them to accept it is time to stop. They are all strangers to each other.

The arrival of a newborn is usually an event of great joy and expectation. But what if all those expectations turn empty? Not life but death has announced itself. And yet, there is still a baby whose birth is real, however still-born. This too is a profoundly ethical moment that touches all those present. This is what those who practice in neonatal-perinatal medicine behold in birth and death. The health professional is the witness of what is most originary of human existence, the most primal of primordiality. Here is an ending that never had a beginning. We wonder what appeal and what touch is experienced in such moments. What does it mean to speak of an ethical experience for a life that was only shortly lived?

It is not merely that ethics infuses or obfuscates morality with meaning. Ethics is fundamental to our existence, an existence that we collectively share with others in life and death. And so, as healthcare professionals, we are vulnerable to the ethical founded in our encounters with newborns and families for whom we care, in spite of being unfamiliar to each other.

> John was very much a wanted child. His parents had tried for years to get pregnant, utilizing all available fertility services. Several times, their attempts ended in early miscarriage, loss after loss, until one pregnancy matured to 24 weeks gestation. It seemed that they were finally going to have a baby. But then contractions developed, evolving into premature labor. The doctors and nurses could not stop it, and signs of fetal distress were evident. The parents knew their options: intervene or await the inevitable. A low transverse incision caesarean section was emergently performed. But it was already too late; the baby lay still at birth. In the aftermath and grief that followed, the parents were told that it would be unwise to get pregnant again given the risks of uterine rupture and other maternal health issues. Yet, the parents persisted in their attempts. Hormone supplements were started. Another pregnancy was achieved. At the parents' request, a prophylactic cervical cerclage was placed, stitching the cervix closed, to try to maintain and support the pregnancy. The parents made it past 20 weeks, beyond the borderline of viability of 22 to 24 weeks, to what many would describe as a healthy gestation of 35 weeks. They bought a car seat, baby clothes, and made all matter of preparations for John's birth. Everything seemed fine. Given the risks of uterine rupture, an attempt of vaginal delivery was precluded. A caesarean section was booked for 37 weeks gestation. But on a fateful night, John's mother awoke to blood-soaked sheets, and although attempts were made to get John out in time, he was not born unscathed. John was seriously sick at birth, needing mechanical ventilation, inotropes, blood product transfusions, and other measures of intensive care. John had suffered severe brain injury. The parents were given the option to discontinue medical treatments. There really was just so much injury. They wanted John to live. He survived. Now John is a few years old, profoundly affected. He is nearly blind and probably deaf. He has no speech. He cannot protect his airway, needing constant suctioning to prevent him from choking on his secretions. All his food is given by gastrostomy tube. He is bedridden. His mother or father sleeps bedside him every night, night after night. There is no question that they love him. They say that they would make all of the same decisions if given the chance all over again. Sometimes they ask whether the caesarean section should have been done earlier, telling

others how John was perfect before he was born. In these moments, we are reminded that the injured child they have is John, but also not John.

Ethics manifests itself as something originary that touches us in the inceptuality of the life of the child and that may determine the course of the remainder of our lives. The ethics that was born with John provides the source, supply, and strength for meeting the challenges that John's parents suffer and counter. We need to acknowledge the intensity and force of this ethics that may appear not simply as a need to respond but as responsibility. When parents encounter the newborn infant it is then possible (even likely) that they undergo a profound transformation: They have been 'touched.' Their touched existence is an ethical transformation: "It cannot be put away, but becomes who one is and one's way of being in the world" (Donohoe, 2012, p. 194). They are now in the spell of this loved newborn, even if such a life is one of challenges.

While the ethical demand is formative of 'me,' the moral presumes agency, making me an 'I.' The moral question asks, how ought 'I' to act? The 'I' reflectively and interpretatively weighs benefits and burdens to deal with issues of justice, society, and so forth. It is this 'I' that judges along the spectrums of right-wrong, good-bad, or just-unjust as possible actions are weighed in their goodness. The 'I' expresses my subjective sense of otherness. Any rational or irrational weighing ultimately distances itself from ethics as an originary event: the event where the other appears but also withdraws as a disclosive eventing (*Ereignis*) of being, as already other to me (Heidegger, 1998/1946). The moral must return to settle, rest, or otherwise be felt by 'me.' Ethics emerges in our relationships with others. It is more than a discipline dealing with what is 'good' or 'bad.' Still, ethics cannot help but be normative in our 'subjective relation' to 'that which is other.' It stirs a response in our-selves, in its 'goodness' or 'badness.'

The inceptual ethics of natality is complex in its beginning: How do ethics come into being as parents seek to conceive a child? How do the ethics of pregnancy differ from ethics encountered after birth? How do the ethics of the past bear on the present and future? The temporality of the ethics of neonatal-perinatal medicine does not simply begin, nor certainly end, with the birth of the child. And yet, because a child is born into the midst of ethics, bearing traces, residues, and memories of the child's conception and pregnancy, practitioners in neonatal-perinatal medicine need to be aware of this metamorphosis.

Everyone knew that Lynn was dying. She had been born with a complex heart disease, total anomalous pulmonary venous return. Initially, the prognosis was deemed as excellent. She was quite well after birth. Her parents were able to spend her first days holding, cuddling, and otherwise spending time with her relatively free of medical technologies. When the day of surgery came, they nervously kissed her forehead as she was brought to the operating room. There, after putting her on cardiopulmonary bypass, the surgeon implanted the anomalous pulmonary veins into the left atrium, creating the anatomical connections that most babies are born with. She was brought back to the intensive care unit on mechanical ventilation and vasopressor support. Her parents were relieved, and they began to talk about home in the coming weeks. Medications were decreased, and breathing support weaned. But then recovery slowed. Lynn could not be taken off the breathing machine. Various tests were completed, x-rays of her lungs, ultrasounds of her diaphragms, laryngoscopic assessments of her airway, and so forth. Without a diagnosis of her breathing-machine dependence, a lung biopsy was discussed. Her family, with patience and understanding, agreed. The family was absolutely loving, always present at her bedside: reading her stories, holding her, and participating in her care. Everyone recognized that they were good parents. When the diagnosis of pulmonary lymphangiectasia (a progressive, irreversible lung disease) was made, the family accepted it with sorrow. They knew that it no longer made sense to continue on with intensive care. They saw what the doctors, nurses, and other medical team members saw. Lynn could not be weaned off the breathing machine, and her lung function was deteriorating. And yet, for her parents, right now was not the time to stop. They did not want her dying over the holidays. They did not want memories of her death to coincide with this time for their family. So, they asked that medical treatments be continued—to carry on for somewhat longer. Surely the request was reasonable. Other babies had been supported for longer periods of time as parents came to terms with such diagnoses. And yet, as the mechanical ventilation was continued, it was apparent that keeping Lynn comfortable necessitated escalating infusions of analgesics, sedatives, and other medications. The medical team was trying to do 'right' by the parents but worried that intensive care was bringing additional hurt and prolonged suffering to Lynn.

The first cry of the child announces the appeal of the natality of ethics followed hopefully by moments of touch, holding, and togetherness. However, the birth of joy is also the birth of pain. Who knows better the stirring of these ethics than those who have lived through neonatal-perinatal medicine? The community of the medical team is professionally tied to the existential ethics of new birth that is received and attended in neonatal care. However, in neonatal-perinatal medicine, it is not solely the children that need care but also their families.

At times, healthcare professionals find themselves in a state of inaction as they worry, struggle, or are otherwise unsure about what action to take that seems appropriate or right. Sometimes, they are immobilized by ethics. Bio-ethicists, philosophers, and practitioners need to realize moral questions (if dependent on the ethical) cannot necessarily be resolved from knowledge of the implications and consequences of our actions. Ethics transcends theories of morality. And so moral theory cannot be employed to address ethics as a fundamental phenomenon. It is precisely because of the ethical that the moral experience of the right-wrong, good-bad, or just-unjust may occur. In these stories, we encounter or discern the primal significance of ethics. Ethics as a primordial phenomenon may touch, affect, or compel us to be and act morally.

For many people, it is difficult to hear or read these stories and not feel moved. And for those with knowledge in neonatal-perinatal medicine, these anecdotes may stir memories of past events, call to mind present situations, or even foreshadow future circumstances. Surely each story is unique, recounting certain particulars of an individual patient, his or her family, and the involved healthcare team members. They are fragments of larger life stories of pregnancy and birth, medical intervention and decision making, life and death. However, each anecdote also conveys equivocal understandings of human experiences. As readers, we sense hope, despair, joy, worry, uncertainty, resolve, loss, love, and other emotions even if such emotions did not arise from our own circumstances. The anecdotes do not convey simple, specific matters of fact but instead stir us sympathetically in their ethics.

Even if none of these stories would necessitate an ethics consult or lead to legal review, they are all profoundly ethical. Ethics permeates the stories of each and every child receiving medical care, even the so-called uncomplicated cases, manifesting something originary that touches us. We need to acknowledge these ethics in the context of medical-technical challenges and innovations that are formative of the practice of neonatal-perinatal medicine. And it is not merely that new tests and treatments have become available for infants born premature, marked by congenital anomalies, or beset with transitional problems (Avery, 1992); the culture of care itself is always changing, responding to a plurality of patient-family and interdisciplinary-professional perspectives (Davis et al., 2003; Hall et al., 2017; Philip, 2005).

Was it 'right' to resuscitate Cody, even without his mother's consent? Does the remote possibility of survival rationalize the prolonged suffering of Mohamed? What features and outcomes should guide decision making for John? Did the healthcare professionals veer from their responsibilities for Lynn by delaying her dying? Who needs to speak for the child who cannot speak for him or herself? What are considered to be the appropriate medical interventions?

Healthcare professionals have to determine how to resolve moral decisions. We need answers for particular patients and specific bio-ethical issues to guide clinical practice, even if such determinations are provisional, fluid, or otherwise under constant consideration for change. As well, what the bio-ethical discourse of neonatal-perinatal medicine also needs to consider is why we ask such questions in the first place.

The aim of this book is to explore the ethical-moral complexities of neonatal-perinatal medicine from a phenomenological perspective. Bio-ethics tends to be understood as a critical normative discourse drawing from moral theory; phenomenology expresses a concern for meaning as it arises in the world within which we live. In comparing bio-ethics to phenomenology, Fox (1990) writes,

> Bioethics proceeds in a largely deductive manner, formalistically applying its mode of reasoning to the phenomenological reality it addresses.
>
> (p. 207)

The point is that bio-ethics and phenomenology do not need to be considered separate disciplines. More so, we need to appreciate how approaching constitutive moral questions, without concern for phenomenology, offers a theoretical discourse lacking in ethics. Values, principles, rights, and so forth cannot 'simplify' ethics for us.

As a philosophical tradition, we identify phenomenology with the work of philosophers such as Edmund Husserl, Martin Heidegger, Emmanuel Levinas, Simone de Beauvoir, Maurice Merleau-Ponty, and Jean-Paul Sartre; and developed further by contemporary figures such as Jean-Luc Nancy, Hubert Dreyfus, Don Ihde, Claude Romano, and Jean-Luc Marion. Although each of these figures varies in the aims and articulations of their work, common to all is a preoccupation with unfolding understandings of life meaning from the reality of life. Here, we find the registrar for a phenomenological ethics.

> To look at ethical involvement from the neutral standpoint of theoretical reasoning is not the right phenomenological way to approach this issue. Rather, the ethical experience itself has to be described in a way that it captures the lived experience of being called, of being addressed by an "ought" itself. In doing this, phenomenological description exceeds the dichotomy between "the descriptive" and "the normative."
>
> *(Loidolt, 2012, p. 14)*

> Let us try to assume our fundamental ambiguity. It is in the knowledge of genuine conditions of our life that we must draw our strength to live and our reason for acting.
>
> *(de Beauvoir, 1976/1948, p. 9)*

Even the most sophisticated of philosophical, phenomenological arguments rest on concrete and often evocative descriptions of human experiences. In a similar yet different way, scholars and researchers in practical disciplines such as education, psychology, nursing, social work, and medicine have been inspired by phenomenological philosophical traditions to explore human experiences and the significances of such experiences (Giorgi, 1970, 2009; van Manen, 1990, 2014a). For professional practice, phenomenological work has the potential to cultivate insights and sensitivity regarding the ways that others experience the world (van Manen, 2014a).

Phenomenological insights and understandings may serve as an integrated or critical perspective on established bio-ethical theory, or contribute in a more fundamental way to understanding the human condition (Svenaeus, 2017). From such a perspective, phenomenological research is always, already ethical, as the aim is to turn outwards to the life experiences of others, recognizing we cannot reduce others' experiences to our own or assume that all others experience the world in a univocal manner. Where do we look for the beginnings of a phenomenological ethics?

This book consists of three sections: before a child is born, the cradle of the neonatal intensive care unit, and ethics and decisions. The sections and chapters blend into the themes of inceptuality and natality of ethics that arise in the conception, birth, and coming into being of the preborn and the newborn. Understanding this fundamental ethics of neonatal-perinatal medicine is crucial for approaching all manner of moral dilemmas. Each section should be read as telling fragments of the larger picture of the ethical-moral complexity of neonatal-perinatal medicine.

Section I, 'Before a Child is Born'—The ethics of pregnancy has received sustained attention by bio-ethicists around such issues as assisted reproduction, abortion, parental rights, germline experimentation, genetic testing, and eugenics, to name a few. At the heart of these topics lies the ethics of the fetus in its becoming. The ethics of the fetus does not merely point to the moral status or 'personhood' of the fetus; but also to the recognition that fetuses are growing in the bodies of women such that the lives of fetuses and the women who carry them are intimately intertwined. From a phenomenological perspective, we need to wonder and ask about how it is that we encounter the ethics of pregnancy. Such understanding is fundamental to any exploration of bio-ethical issues.

Section II, 'In the Cradle of the Newborn Intensive Care Unit'—This section explores the experience of ethical responsivity from the perspective of the parent whose child requires medical care. From a parental point of view, ethical responsibility for the child is not necessarily due to a formal code of ethics. Parental responsibility arises in the encounter of the mother or father with the newborn. What does it mean for a parent to experience ethical responsibility for his or her child? And, how is this responsibility phenomenologically associated with responsivity to the newborn? To develop an understanding of parental responsivity to the newborn in the context of neonatal-perinatal care, we need to be attentive to the empirical variety of experiential encounters that may occur in this medical setting. The concern is with the lived meaning of ethics itself as it originates and wells up in the parent's experience of being touched by his or her child. Examples are taken from the context of neonatal intensive care, whereby the condition of the child and the technomedical environment have the potential of complicating contact of parent and child.

Section III, 'Ethics and Decisions'—When we speak of ethics in medicine, we tend to think of the moral correctness of specific decisions and actions. We may question if starting or stopping a particular intervention is right or wrong. At times, we may even wonder whether providing any medical treatment is appropriate. In neonatal-perinatal medicine, the focal concern of bio-ethical deliberations for the medical staff is usually the newborn child, with the parent carrying substitutive responsibility in moral decision making. This responsibility is in part a jurisprudential matter, written into the professional moral code of medical practice and the administrative institution within which health professionals operate. This section explores a manifold of ways moral decisions may be conducted and experienced.

Phenomenological studies can never be complete, as we recognize our existence in the world is an interminable source of meaning. This work may be judged as successful if it stimulates the reader's wondering, thoughtfulness, and sensitivity to the ethics of neonatal-perinatal medicine. I hope that this book not only serves healthcare professionals concerned with the ethics of clinical practice in neonatal-perinatal medicine but also that parents read it as well. While healthcare professionals in neonatal-perinatal medicine must be able to ask the question of what kind of ethics are appropriate sources for thinking and acting in ethically-sensitive situations, families ultimately face, make, and live with the consequences of the moral decisions affecting their children. Bio-ethics should not simply be considered as a system of rules that are applied to reach moral decisions. The philosopher Emmanuel Levinas (1969/1961) shows us that we experience the ethical in the encounter with the other before we are involved in bio-ethical thinking, reflecting, and reasoning.

> ethics is an optics ... it is a "vision" without image, bereft of the synoptic and totalizing objectifying virtues of vision; a relation of an intentionality of a wholly different type.
>
> *(Levinas, 1969/1961, p. 23)*

Bio-ethics adds value to practice by supporting reflection and understanding of the meaning of actions and inactions taken in the clinical practice.

Finally, this text should be read as a reminder that in neonatal-perinatal medicine, ethics exist not only in the setting of moral complexities and uncertainties inherent in clinical situations but also in the day-to-day encounters of healthcare professionals with children and their families. The value of phenomenology for neonatal-perinatal medicine emphasizes a return to experience as lived. From a foundational perspective, phenomenological research gives the questioning of the meaning of a phenomenon priority over and above gaining truths, drawing conclusions, or establishing a theory. This questioning may contribute to new knowledge by opening a way of seeing, and a way of questioning, that would otherwise be too readily taken for granted. So, let us reflect on the ethics of neonatal-perinatal medicine.

SECTION I
Before a Child is Born

CONCEIVING ETHICS

"I think we can start trying." The conversation began just like that. We had stable jobs, a mortgage that we were beginning to pay down. We both seemed healthy. There did not seem to be a reason to wait. Then again, some things would have to change. There was parental leave, and after, one of us would probably need to cut back on our work hours, at least for the first year. Perhaps we ought to think about a bigger car? Would there still be enough time for our hobbies? Should we risk our present quality of life? Everything is going so well right now. So, there it was: the question. Should we? It was probably one of the most significant decisions we would ever make. Should we have a baby?

When couples talk about the possibility of having a child, all manner of thoughts, considerations, and reflections may be articulated: Does having a child fit with our life? Are we going to be happy as parents? Will a baby make our life more meaningful for us? Surely, a great deal of rationalizing can compose these discussions: approaching, testing, and encircling the question of whether or not to have a baby. However, despite the 'object' of each question being the baby—with the 'subject' (the parent) expressed as 'I' or 'we'—a regard for what is 'good' for this future child may appear absent. It is as if the parents contemplate creating the baby for themselves or with primary consideration for the impact the child will have on their lives. However, in asking the question should 'we' have a baby, do the prospective parents not take ownership, express accountability for the possible existence of a child? Is this not where ethics is conceived?

Travis Rieder points out that when we conceive a new life we "inflict the child on the world, and inflict the world onto the child" (Beard, 2018). While the term 'inflict' does not necessarily reflect the sensibility in which we want to think about having a child, this quote points to the ethics that are missing from egocentric questions that fail to reflect on the life of the child. How is our responsibility founded in an imagined sense of him or her?

The ethics of conception is not something we necessarily think about. Many people go through their daily lives with little thought of having children. Life is already full, busy, occupied by other matters such that when conception 'just happens' it is unexpected. For others, having a child exists perhaps as a future possibility or was a past consideration, but in this particular moment of 'now,' pregnancy is something that needs to be guarded against by decisions of abstinence, methods of birth control, or simply living life in such a way that it is an impossibility. The preference is to be 'childless' or 'childfree.' But then, there may come a time in people's lives when the idea of having a child comes into being.

> After my husband and I got married and felt some semblance of being settled in our new life together, it seemed only natural that getting pregnant was the next step. We never had a specific conversation, but I do remember the day, standing in the bathroom, holding a contraceptive pill in my hand, rolling it between my fingers before resolutely putting it down. "We are going to have a baby," I said out loud to myself. Downstairs, in the kitchen, we were making our coffee and getting ready for the day. Between bites of toast, I just said it rather matter-of-factly, "So I am not going to take the pill anymore, and we will just see what happens."

For some, the idea of having a child was always there, anticipated from childhood, that just like our parents we too would become parents one day. It is as if the thought of procreation is in itself inborn. We expect it of ourselves, or others expect it of us. Perhaps siblings, friends, colleagues, and others all seem to be having children—everyone else is doing it. So, when conditions seem fertile, the idea takes hold of us to unfold, as we feel the pull to have a child of our own. Our lives are somehow now ready enough; having a child becomes a possibility. Still we may wonder, what is the ethics of having a child? Who might this child be?

> So, I am waiting in line at the grocery market when a baby in a nearby stroller catches my eye. She seems to notice my looking, staring back at me. Whatever was on my mind is gone as I find myself wondering in a daydream: If we had a child, would it be a girl? Would her hair be brown like mine? And skin so fair? I would not dress her in such dresses. And certainly would not put such dainty things in her hair. "Cash or credit?" the cashier interrupts.

Husserl (2005) describes the experience of imagining in the mode of 'inactuality.' For the imagined (inactual) child-to-be, even though there is no actual child, my imaginary experience still has sense content. However, the content is not the same as it would be if it were present: "it is as though it were there, but only as though" (p. 1984). We may say there is an indistinctness, vagueness, or otherwise tacit quality to what is present yet absent. We appreciate that the parent may dreamily envision cradling a child, pushing a stroller, changing diapers, and so forth. And yet, even in imagination the child does not exist apart from us. It is, as Simone de

Beauvoir (2004/1945) writes, "... impossible to define an object in cutting it off from the subject through which and for which it is object; and the subject reveals itself only through the objects in which it is engaged" (p. 160). When this phantasy is interrupted by our actual surroundings, the imagined child remains a possibility in 'our' lives.

Even before conception, we may begin to try out names, purchase infant clothing, buy a car seat, or otherwise make room for a child in the relational, imagined reality of a future life as a parent. We place the child at home in our thoughts as if he or she has already come into being. Boström (2016) writes of "making room" in the dwelling of our lives as one of our first acts of becoming a parent.

> There are no clear limits to how early the process of making room can start or how intense and serious it can get in earlier phases, although there might of course be personal and cultural limitations on how real the unborn child can become to us.
>
> *(p. 148)*

We create a child even before its biological beginnings through imagining its potential existence within this inactuality to become real to us (Boström, 2016, p. 156). And yet, the ethics of this inceptuality, of this imagined child, cannot help but be paradoxical. On the one hand, we only find out about pregnancy after we have already conceived a child, meaning the child as biologically existing is only recognized after it has already come into existence. While on the other hand, the expected child may already be experienced as a real absent presence.

> "When are you going to get on to having some children?" It was a perfectly reasonable question to ask, especially considering we had been married for four years, and I was turning 38. I replied, "Maybe we'll start trying next year." But the truth was we had been trying for almost three years. We had both had testing done. Nothing was physically wrong with either of us. It was just not happening for us, and perhaps it never will. I do not even know why I want to have children. It's just a missing part of our life that others seem to have. I tried to keep these thoughts hidden from my expression, turning away.

In its absence, the imagined child is present. Sartre wrote that that which is "imagined is a *definite lack;* it stands out as a cavity" (2004/1940, p. 126). In the image of a cavity, we find the relational void of a parent who experiences a wanting lack. When a parent imagines an absent child, he or she may experience the child in its "irreality" (p. 159), such that the parent experiences his or her actual world in terms of the child's absence: what is 'irreal' as being 'real.'

Some couples can get pregnant with seemingly minimal effort. It may just happen after missed or failed contraception, or when the possibility of a child is left to chance. But for others, this is not the case. Reported fertility rates are at an all-time low (Martin et al., 2018; Milan, 2014). When the idea of pregnancy comes into being—to have a baby, to become a parent—we realize that an ethics too has already come into existence.

> We want to have a healthy child. I do not care at this point about the risks of it all. Do not get me wrong. The miscarriages have been heartbreaking, and I hope to God to never have another baby born at 23 weeks again. Having Arianna, trying to keep her alive, only to lose her, to have her die days after she was born in the NICU—I still struggle with whether putting her through all of that was right. But, I want to have a child. I want to grow our family. If I can accept the risks, surely others can as well.

Some parents deeply ponder the ethics of conceiving a child, whether they can provide a life for him or her, or even whether this is the kind of world into which they want to bring a child. Others consider whether they want to chance a risky pregnancy: ending in a premature birth, to endure medical treatments, to suffer medical interventions, or even to die. In such thinking we find an imagination that permeates all thought and conceptualization (Ricoeur, 1977/1975). This new life may stir a response: "An expectant parent is moved by the power of a beginning life full of possibilities and newness. The infant ushers in absolute newness, a life to be lived unlike any life lived so far" (Donohoe, 2012, p. 188).

Still, we need to remember that the imagined child of conception has a particular ambiguity, equivocality, and absence such that his or her existence while being meaningful (in coming into existence) does not eclipse other significances and meanings founded in the expectations of pregnancy.

> We conceived in a month, reached 12 weeks, and we were starting to tell people. It was like we had joined an exclusive club. I found myself nodding and smiling knowingly at new parents as they pushed strollers whenever I walked by them in grocery stores, the mall, and other public spaces. Even though they were all strangers to me, we had some shared connection. I was completely different than I had been just months ago. But then, a week later, I woke up to my sheets soaked with blood: The pregnancy was over. We never knew if it was a girl or a boy. Surely, we mourned it, the baby. But that was not all we lost.

We know that people may desire pregnancy for all manner of different reasons beyond the obvious wanting to have a child. Some get pregnant at the behest of their partner, to attempt to save a failing relationship (or even establish a relationship in the first place), to fit in with perceived social norms, to fulfill some personal sense of existential void, or to establish a legacy. We realize such meanings not only from the motives that people voice when they recount their rationales for having a baby but also from what people express as lost when a pregnancy ends.

We also need to acknowledge that, for some, pregnancy is not a choice one can freely make. Many women lack reproductive rights let alone basic autonomy when it comes to decision making about their bodies. However, in societies and situations where pregnancy can truly be a choice, a choice for the parents, we need to wonder: How is having a child ultimately for the sake of the child in itself? Alternatively, if we accept that the child as an imagined presence is brought into being by the parent, perhaps conception is truly not for him or her? After all, the beginnings of pregnancy may find its inception in response to motives that are relatively distant from, or intertwined with, a felt responsibility for an imagined child.

> We tried intrauterine insemination, and eventually—three years after we started trying—we pursued IVF. Despite being on supplements, vitamins, hormones, and other medications, the first two rounds were unsuccessful. All our friends had families, and we wanted one too. We so wanted the next round to take. So, they put in two embryos, and I conceived: twins. I finally had got one to stick. We never really wanted two children. But we will take what we can get.

With medical innovation, at times we can overcome infertility. Many different assisted reproductive technologies are available: ovulation induction, in vitro fertilization (IVF), artificial insemination, intracytoplasmic sperm injection, and so forth. And yet, achieving pregnancy with such means poses risks. Multiple gestation pregnancies (twin, triplet, and so forth) while being associated with assisted reproductive technologies are also associated with preterm birth that carries its own complications (Kushnir et al., 2017; Sunderam et al., 2019). Even when precautions are taken to mitigate the risks of conceiving a multiple gestation pregnancy (such as single- rather than multiple-embryo transfer), preterm birth and congenital anomalies remain present as risks (Liberman et al., 2017). In other words, assisted reproduction would seem to demand we accept risks for a child even before we conceive him or her. What worry should we have when we knowingly place a child at risk in our efforts to conceive him or her? Should this future child not be first and foremost in our mind even before attempts at conception? Is this not the ethical imperative of conception?

In our postmodern world, many will advocate that children deserve to be born for the right reasons; that they are deserving of love, stability, and so forth from families who can take care of them. And certainly, most children born premature or needing medical attention are conceived 'naturally.' As a society, we consider that families that are prepared to provide for children have the 'right' to be supported to have them—even when such technomedical efforts implicitly pose risks to the child. In comparison, having children as a source for family labor, funds for dowry, or otherwise reducing them to a means of secondary gain is considered morally wrong.

Still such deliberative moral reasoning may pass over the nuanced complexity of conception ethics, whereby the existence of the imagined child has a particular ambiguity, equivocality, and absence that may at times seem distant from such decisions. And of course, quite plainly we need to acknowledge that, by and large, we support people to have children simply because they want them. For some parents who so desperately want a child we rationalize whatever means to support the conception of a pregnancy.

> We tried IVF, but they could not stimulate my ovaries to release any eggs. We started thinking about donors, reading on the internet, looking locally and even abroad. We wanted the baby to be ours, and it just seemed in Canada that there would always be some connection to the donor, like it was some adoption. So, we found a clinic abroad, paid for all of it up front. We did not know much about the donor except she was also Caucasian, had hair like me, and was tall like me. The baby would look like me. It would be mine. Todd gave his sperm sample, and I was implanted. We came home and waited. After waiting for 2 weeks, I picked up a pregnancy test. A faint line appeared. My baby was in there.

As a starting point for exploring the ethics of neonatal–perinatal medicine, we realize that the ethics of conception may in part be founded in an ambiguous imagined reality. In other words, before conception we may create the child as an other for our selves, whose very existence and moral worth is founded in his or her envisioned relation to us, as we imagine his or her future and how he or she will fit into our family. And while a child conceived by natural or artificial means may be wanted, and that in itself most people would likely support as a sound reason to have a child, we also need to acknowledge the nuanced complexity of the ethics of an imagined child. The possibility exists that imagination will not equal reality.

PREGNANT WITH CHILD

We met with the doctor and talked about what it meant and what could be done. It seemed like something fixable. Still, the doctor said it was early enough on in the pregnancy that we could chose to terminate. It was my choice whether I wanted to have a child who required heart surgeries and potentially so much other medical care. I remember going home that evening and lying in bed with my hands on my belly. I tried to feel her. She was in there, but felt distant. I really wanted to have a child, but I was not sure about this.

Pregnancy names the situation of a mother; it is from the Latin *praegnantem*, being "with child" (OED). It is the time of first impressions, first touches, first bodily feelings of 'who' will be born as a newborn. And yet, both in the lay and medical literature, we may be reluctant to use the term 'child' to describe the life that is growing within. The proper term to describe the child-to-be is 'fetus,' relegating antiquated or old-fashioned phrases such as 'with child.' In comparison, from a medical perspective, we frequently describe pregnancy as a condition or state, conceptualizing it as an ailment, disturbance, or otherwise affected existence at risk of possible complications. But where is the child in this discourse? Does our contemporary language betray an awareness that 'who is inside' is indeed not properly a child? How does pregnancy give the being of this yet-to-be born child ethically to the mother? And, how does pregnancy give the child ethically to the social world of others?

Embryology tells us that pregnancy follows conception: The union of sperm with an ovum creates a zygote that matures over the course of days to implant in the uterine wall. Through cell growth, division, and differentiation, embryogenesis occurs as these embryotic clustering of cells mature to be called a fetus by nine weeks' gestation. A brief opening of any embryology textbook will reveal that there is indeed much going on. Accordingly, it is without surprise that embryology has been divided into complex fields of descriptive, comparative, evolutionary and teratology study. And yet, for the first few days to weeks to months, the baby's presence may not be evident to anyone, including the mother herself. Instead, in the beginning, pregnancy can only be 'found out.'

> My period was late, not super late, but late enough. It took me a few days to convince myself that it was not going to be just one of those irregular, heavy ones. I needed to buy a test. I was just driving myself crazy. It took me time to somehow accept that there was a chance the test could be positive. So, when I finally bought it, it was not until the evening. I was alone, standing in the bathroom, waiting for the result. A plus sign. I did not feel pregnant, but when that plus sign appeared, I stood in front of the mirror pulling up my shirt to expose my belly bare. Was I different?

Surely the beginnings of pregnancy can be surreal—"I just need to get used to the idea of it" or "I am just sorting out how I feel about it"—as if there is an even- tuality to pregnancy that the test alone cannot give. Even if the test is a clear positive, we would not be surprised to hear an expectant parent say, "This cannot be real. Am I pregnant?"

Pregnancy heralds more changes than a missed period. The feminist philosopher Iris Marion Young writes:

> As my pregnancy begins, I experience it as a change in my body; I become different from what I have been. My nipples become reddened and tender; my belly swells into a pear. I feel this elastic around my waist, itching, this round, hard middle replacing the doughy belly with which I still identify.
>
> *(Young, 1984, p. 49)*

The inclination or temptation of the healthcare professionals may be to assume that such descriptions should be read to interpret the experience of pregnancy as a constellation of physical symptoms: fatigue, nausea, mastalgia, edema, pruritus, polyuria, and so forth. However, to be pregnant transcends the body as merely symptomatic. The pregnant body is a transformative body, a changing body, bearing a growing child within.

> I noticed my body change first: old pants becoming too tight, familiar foods tasting strange, and, of course, just being tired. And then it happened, a few months in, fluttering feelings in my belly. I was lying in bed, just resting on my back, waiting to fall asleep when I felt something. It was almost like gas, like bubbles popping deep within my tummy. I shifted my body without even thinking, as if I could make those feelings go away, only to notice them again. I could not say whether they were kicks, punches, or simply her moving around. Surely, they were feelings of my body, but they were coming from someone other than me.

As Pugliese writes, "A woman finds herself pregnant even though she may have planned and prepared for it" (2016, p. 72). Even if conception itself was the results of desire and deliberate action, pregnancy can only be discovered after it has come to be. The *quickening*—the perceived movements of the fetus—expresses the meaning of 'quick,' from the Old English *cwic*, as "living, alive, animate" (OED). It is perhaps not surprising that the fetal movements described as quickening are aligned to the moral-legal history of feticide and abortion (Dunstan, 1984). When a woman discovers herself to be 'with child,' she comes to an experiential awareness that within her changing body is an other.

> And then it happened. I was driving home from work, caught in traffic, when I felt a painless but unmistakable punch in the gut. My hand went to my tummy reactively. These were the first movements of my child. I just held my hand there waiting, feeling to feel.

Pregnancy expresses a dynamic *ethos*, a relation of touch and responsiveness that marks a changing bodily entangling of mother and child: "a temporality of movement, growth, and change" (Young, 1984, p. 49). And through this becoming, "we create the unborn child by acknowledging that it exists, that it hears us, that it has a name, feelings and responses: the it becomes an *other* by being treated as a *subject*" (Boström, 2016, p. 151).

> He was just not letting me lay on my side. Every time I would seem to get comfortable on the couch, he kicked, disturbing me. He has a mind of his own, and I am the only one who truly knows it.

For the mother, we can appreciate that she has a "privileged relation" (Young, 1984, p. 46). The partner cannot quite know this yet-to-be-born child is this way. Mothers, after all, are "tied to experiencing the baby through their own bodies" (Bergum, 1997, p. 145). And yet, the otherness of the baby may transcend the mother's bodily being.

> She experiences her body as herself and not herself. Its inner movements belong to another being, yet they are not other, because her body boundaries shift and because her bodily self-location is focused on her trunk in addition to her head.
>
> *(Young, 1984, p. 46)*

It is not merely that this relation is privileged and dynamic; it is inceptual in its becoming. Mumford (2013) writes: "The commencement of that relationship, the moment of first encounter, the moment of creation, is not observable, not given to know in experience" (p. 25). Instead, the movements of this yet-to-be-born child reveal the expectant mother to have already been in bodily relation with her child. We might say pregnancy expresses a constant discovery of the child in its development, already having been present. At the same time, we need to be careful in "underestimating the individual differences between pregnancies" (Svenaeus, 2017, p. 111).

> I never felt butterflies, flutters, or any of those airy things people describe as first feelings of pregnancy. It was only later that I felt kicks, stretches, and other movements, not that unlike someone touching my tummy, from the outside. There would be times where I would feel a poke, and then I would poke back. It was like an addictive game of exchanging pokes, like I was caught in a private game with my child. Other times, I might poke without thinking, perhaps unconsciously, and then feel for the response. I became so accustomed to this routine that to poke and not get a response I can only imagine would somehow have left me feeling strange.

We recognize in pregnancy that the yet-to-be-born child could not be nearer to the mother: She harbors the child within her. Moments of touch and touching may become so routine that without them the mother's pregnant body itself may become unfamiliar. Still, although the child is continually present, growing and developing in the woman's changing body, the child is not necessarily felt continuously as an other to the mother.

> For much of my pregnancy, I did not have a child on my mind. It was more the massiveness of my ever-expanding body. I was just in a condition of constantly carrying my belly around. Near the end of it, I was looking for chairs to sit, elevators to avoid stairs. The reflux, the bloating, the constant having to just go pee, by the last months of pregnancy, I was just done.

At times, the physical manifestations of pregnancy may overshadow feelings of otherness within. And as the pregnancy progresses, we can appreciate how the mother may feel ready for the pregnancy to be over: She desires the child to quite simply come out, to end the so-called physical symptoms of pregnancy (yet also foreseeably anticipate meeting her expected child, outside of her body).

As for the partner, the child may appear obliquely and outwardly as the partner experiences the becoming of the child through the changes in body size and shape of the pregnant mother. A well-placed sensitive hand may feel for the unborn child. Phenomenologically, the hand experiences a double touching: waiting to be touched by the child from within and touching the child in response. It is as if the child exists in an in-appropriable 'other' world, a 'womb world' (van Manen, 2019). The partner only has access to feeling this developing life by contact with and through her. Buytendijk observes that even more than the human voice and the look, the touch strikes us in a particular intimate way. It is "a fundamental characteristic of the touch, that it obtains in inter-human relationships the meaning of an existential communication" and of intimacy (1970, p. 115). So, we can recognize the possible difference in felt intimacy and felt distance that may pervade the relation of the partner with the unborn child compared to the mother.

In these possibly phenomenally different situations, we need to acknowledge the natality of the ethics of neonatal-perinatal medicine. Ethics unfolds as this 'other,' this yet-to-be-born child is dynamically and idiosyncratically given in relation. And we may wonder about the experiential equivalents of the subtle signs that give the child in expectation.

> Although I am now five months into my pregnancy, the realization that we are having a child is still surreal. We found out that it is a girl and have already chosen a name. And although I use that name to name her when she is kicking, waking me up in the middle of the night, I still wonder if that name is going to fit.

Inhering even in the intimate, bodily relation of an expectant mother with child is ambiguity. There is an otherness to the child that remains difficult to name, experienced beyond expectation. We can understand how for some parents it is not merely culturally, socially, or preferentially untimely to give their still unborn child a 'name'—it is experientially inappropriate because the child has not yet come fully into being as a presence. The yet-to-be-born child truly has 'not yet' been born, transcending expectation. Etymologically, the word 'expect' derives from the confluence of *exas*, "thoroughly," and *spectare*, "to look" (OED). To expect is to regard someone (or something) as likely to fulfill an established view or match with a known image. The act of naming brings the child into being, but the named child may not be the child who comes to be born.

It is significant for bio-ethics how we encounter this yet-to-be-born child, how we find ourselves responsible for him or her, as the encounter shapes moral discussions on life's beginnings and perinatal decision making. Responsibility here does not merely mean the accountability of a subject, meaning how an individual bears culpability or commendation for his or her actions. Instead, the term 'responsibility' points to the *ethos* of living with others, whereby we may find ourselves stirred, drawn, or otherwise fundamentally responsive (Raffoul, 2010). The German term for responsibility, *Verantwortung*, used by Jonas (1984) and Habermas (2003/2001), captures this understanding of responsibility perhaps more fully than its English equivalent through its sense of 'answering' (*Antworten*) to an other.

From this tentative phenomenological exploration of pregnancy, we need to acknowledge that the relation of parent and child is inceptual in pregnancy. The temporality of pregnancy, including the bodily changes of mother and felt changes of the child, instills the fetus with life such that as pregnancy progresses so does the actuality of this fetal life as a child. And yet, we recognize that different individuals may experience this coming into being of the fetus as a child in a myriad of ways. There is no homogenous experience of pregnancy or this child-to-be.

We can hopefully appreciate how decision making for a child before birth cannot help but be clouded by ambiguity and uncertainty, not merely by the ephemeral nature of the decisions under consideration but also through the incomplete, unfolding relation of parent and child. We need to be considerate of understandings we presume, the language we use, and the decisions we make for the child who has yet to be born. The ethics of the pregnancy do not merely point to the moral status or 'personhood' of the fetus but also to the recognition that fetuses are growing within the bodies of women, such that the lives of fetuses and the women who carry them are intimately intertwined (Mumford, 2013). Feminist researchers have reminded us that we need to continue to be attentive to the mother when the medical eye comes to focus on the fetuses with medical issues. The mother ought not to be made invisible, with her needs simply sublimated to those of the fetus or with the ease by which healthcare professionals may provide medical care (De Vries, 2017).

ULTRASOUND IMAGING AND VIRTUAL PRESENCE

Screening ultrasound is the common medical practice of prenatal imaging using sound waves to 'look through' the pregnant body. This imaging is routinely offered to pregnant women in North America, Europe, and elsewhere in the world as part of normal obstetrical care. While early scans may be conducted to assess viability or confirm dates, it is the ultrasound exam centered at 18 to 20 weeks gestation that is performed to look for progressions and potential anomalies. A mother recounts her experience of this medical procedure:

> The technician and I looked at the screen while she was sliding the probe across my belly. I knew that this was supposed to be one of those magic moments of pregnancy, a chance to see my baby. However, I could not really make out the contours of the baby that the technician was pointing at. She was pleasantly chatty: "Here is the head. Here the legs. You can see the hands next to the face. Do you see? Oh, wait a minute …" Then the technician turned silent. I realized that she saw something, and I am not really sure that she knew what she was looking at. She just kept taking more pictures. I kept staring at the last still picture on the screen: a dark blurry image. Is that my baby? Next, without saying anything, the technician left the room, and shortly after came back in again, only to take more and more pictures before leaving and returning again.
>
> I was totally confused, as she said, "You can get dressed now; your doctor will discuss the results with you. He should have the report by tomorrow. Phone his office. He should be able to see you right away for something like this."
>
> I walked back to my car outside the hospital. What could I tell my husband? How would I be able to sleep tonight? Already waiting for the doctor tomorrow seemed like an eternity. As I opened the car door, I cried uncontrollably.

We can recognize that the account of this pregnant woman is unique in two senses. It is unique in the sense that it is the experience of this particular woman. Every experience always uniquely belongs to a particular person. And it is phenomenologically unique in that it differs essentially from other medical encounters employing diagnostic tests.

We may be touched by the psychological, emotional nature of the story. Quite apart from the theoretical bio-ethical questions, the woman who tells her story must be listened to. It turns out that her ultrasound experience is far from 'magical,' as she had hoped. The story speaks to expectations, confusions, anxieties, uncertainties, disappointments, emotions, and stresses experienced by this particular pregnant woman in this specific clinic. The experiential account of the expectant mother offers several potential topics for phenomenological reflection: First, there is the primal experience of the ultrasound technology. The ultrasound lets one see on the screen the preborn infant inside the womb of the mother. What is it like to have this audio-visual technology-mediated experience? Second, there is the experience of diagnosis. What is it like to experience a (protracted) diagnostic judgment? And, third, there is the phenomenology of waiting, waiting for a medical diagnostic outcome. What kind of waiting is this? Let us focus on the unique manner of which routine antenatal audio-visual technologies may shape a parent's experience of their yet-to-be-born child.

> How great is it to see my baby! Honestly, I really didn't believe it was there until I saw its little heartbeat. I cried. It was amazing! The sound of the heartbeat filled the room, and I felt the rhythm to the core of my soul.

For the expectant parent, medical technologies may provide proof that a baby is in there! The audio of the scan can fill the room with the sound of a baby's heartbeat for all to hear. And of course, the ultrasound imaging offers the first possibility for the parents to actually 'see' or perceive the screen image of their child. The ultrasound operator may even be able to discern whether it is a boy or a girl, and then share this information at the parent's optional choice. The phenomenologist Peter-Paul Verbeek (2008) describes the modality in which an ultrasound shows the fetus as individuating. Whereas the mother's womb and the fetus are still integrated in their physiology, the ultrasound breaks through this unity:

> Ultrasound imaging constitutes the fetus as an *individual person;* it is made present as a separate living being, rather than forming a unity with its mother, in whose body it is growing.
>
> *(p. 16)*

Verbeek suggests that the phenomenology of seeing the child's image inside the womb has an experiential effect on the relation between pregnant mother and gestating child: they become individuated and separated while still biologically entwined. We may wonder what maternal knowledge is obfuscated by the technical mediation of the ultrasound. A mother might wonder, somewhat amusedly, whether Verbeek's proposal that the ultrasound individuates the mother and child may be partly given by the fact that Verbeek is a male and would not have felt the undeniable presence of a growing child in his own body. Recall that some mothers interact with their child-within, playfully poking each other. Certainly, there are experiences of individuation in the pregnant body experience of the mother to be, apart from the ultrasound. And even the expecting father may have the experience of hearing and feeling the individuality of the child-within by putting his ear to the belly of the pregnant mother.

Still we can say that for the medical professional the mother becomes an *environment* and the infant a *patient* by virtue of the mediation of the medical ultrasound technology. And yet, it is not merely about what the ultrasound possibly shows; it also absents possibilities.

> We never found out the gender of our first child because we wanted it to be a surprise, but for the second we thought it might be better to know. Part of us was hoping for a girl. I remember the ultrasound technician's words, "It looks like a boy." In that moment of hearing the gender, we had a new picture of what the next years of our life would look like. The girl names disappeared. Imagined images of dresses, dolls, and alike were gone. We were having another boy. And from those words, "It looks like a boy," he became even more real.

We recognize that even in the absence of illness or concern, the activity of ultrasound imaging brings forth characteristics of the child, resolving certain ambiguity inhering in the pregnant relation of expectant parents with child. And we realize, too, that ultrasound also dissipates expectations. In the case of a boy, ultrasound makes 'it' a 'he,' and the possibility of 'she' is gone.

Ironically, the obtained fetal images may appear stereotypical rather than necessarily identifiable as the parents' child in his or her uniqueness. For example, the popularized image of the fetus, often described as a child's 'first photograph,' is far from a detailed likeness of the fetus' actual face. Instead, two-dimensional ultrasound produces the fetal face as a profile, an outline of scalp, forehead, nose, mouth, and chin, complemented by the shadowed features of the brain, tongue, and other interior structures. Still, the pregnant mother may have a hard time making out the infant on the screen.

> When we saw the first ultrasound image we needed help to tell what we were looking at. On the screen, the ultrasound technician pointed out the arms and legs. She was able to get a good profile of the head so we could make out the nose, chin, and mouth. The heart was easy to see as well because of its motion. Those were the parts I remember seeing. The technician could not tell the gender on the scan so it was unclear if it was a boy or girl. Still, we felt

relieved to know that everything looked okay and that we could anticipate a healthy child.

For the expectant mother, discerning the fetal details of the ultrasound image is not necessarily easy or intuitive. Shaped by the processing of the echoing pulses of high-frequency sound waves, the picture on the screen may appear as resembling a snowstorm or a haze to the untrained eye. And so, when something unexpected is observed, as in the opening account of the mother above, we realize that there is more to this technological mediation.

The phenomenological insight is that although the ultrasound individuates the fetus—brings forth by making visible the fetus as a child—the technical experience of ultrasound imaging turns out to be relational in an enigmatic sense. Ultrasound imaging is an experience of *someone* rather than *something* looking inward such that the experience of ultrasound imaging is also an experience of seeing the sonographer and indirectly the presence of the child (who may be healthy or in trouble). When an anomaly or other medical issue is identified, ultrasound imaging is revealed in its sociality, temporality, and diagnostic complexity.

> The technician, she had a look about her, and I don't know, you could just, it was just, it was strange. Everything was normal, we were talking, and then she got quiet. They did not tell me right then and there what they saw. I knew then and there. Something was wrong with my child. The technician knew it and I knew it.

The ultrasound experience may also be the experience of a certain kind of medical diagnosis. The term *diagnose* means "to know thoroughly, to distinguish, discern, determine" (Klein, 1971, p. 209). Ironically perhaps, an ultrasound diagnosis involves seeing through the body to the interior parts. While in the above situation the technical procedure of the ultrasound was not meant to convey a formal diagnosis yet (from the point of view of a hospital or clinic policy); nevertheless, the mother experienced the procedure as wanting to know (diagnostic). She may well have said, "they did not tell me anything." The phenomenology of medical diagnosis is a modern and technological kind of experience that underscores the highly developed technological nature of modern medicine. A diagnosis yields 'knowledge,' and the patient wants to 'know' with some certainty what may be wrong or abnormal. Indeed, the overriding concern of most new mothers (and fathers) is that the baby is all right.

People may respond to waiting for the diagnosis of what is unexpected in a myriad of ways: anxiety, fear, confusion, denial, or indifference. And a diagnosis may be given with more or less certainty, understanding, and sensitivity. Regardless of a patient's emotive response and how he or she experiences waiting for medical judgment, the essence or unique meaning of 'waiting' as a phenomenon or event is not dependent on emotional reactions. Waiting for a medical outcome is a phenomenological topic in its own right. Another phenomenological aspect of the above lived experience descriptions is 'waiting' for the diagnostic result. Waiting is a temporal experience. We may experience several kinds of waiting: waiting for the elevator is different from waiting for a flower to grow; waiting for the rain to replenish the garden is different from waiting for a child to learn to tie a shoelace. Some kinds of waiting can be calculated, while other waiting is subject to contingent and chance factors. We can influence waiting for a child by helping and teaching the child how to tie a shoelace. What kind of phenomenology of waiting is part of the ultrasound experience—waiting to speak with someone who has consequential knowledge about your child?

> This was not our first medical meeting to talk about our baby, but it certainly felt the biggest. In the room were doctors, nurses, and social workers. Joining by way of a video call were people from two other hospitals. It was like we were all sitting at one big table. The doctors took turns talking. First was the talk about the pregnancy, then was the talk about the heart problem. They were worried about other problems, too, with the lungs, so they spent a lot of time talking about what may or may not happen in the delivery room, whether the baby would stay at the birth hospital or need to be transferred somewhere else. We left that meeting knowing there was a plan, and things could ultimately be fixed.

It is not uncommon for healthcare professionals to focus pregnancy talk around medical issues. After all, consultations and conversations are often spurred by a clinical concern. This is not necessarily an altogether regrettable thing but rather expresses a particular attitude of relating to a child. For example, when a medical emergency unfolds, we do not focus on the childness of that particular child but rather the airway, the breathing, the circulation, and so forth. The language of a medical attitude includes: issues, problems, feasibility, and treatment. Subjectivity is left off to the side.

But when the conversational focus is on an isolated part (e.g., congenital heart disease), this medical attitude focuses discussions on the pathophysiology of a child, placing other perspectives in abeyance. The bio-ethical question that needs to be asked is what happens in decision making when we rely on parents to shift from a medical perspective to one which is more holistic—that considers the child in his or her child-ness. Part of this imperative is to support parents shift to appreciate the significance of medical decisions for their yet-to-be born child as a child, that so-called 'best interest' discussions need to consider treatments from a perspective that is broader than their clinical feasibility.

For the practice of neonatal-perinatal medicine, we need to realize that medical technologies disclose the child. A phenomenology of ultrasound aims to understand how ultrasound is experienced in its meaningful, relational, technological and unique aspects; this is also true for a phenomenology of waiting, and for a phenomenology of diagnosis. Showing parents an image of their expectant child cannot help but bring the child to presence in his or her separateness. After all, it is mainly the child, and much less the body of the pregnant woman, displayed on the ultrasound screen.

From a bio-ethics perspective, we could also say that ultrasound reveals the openness of the relation of expectant parent to child in pregnancy. Ultrasound is not necessarily an alienating experience for parents so much as a possible way of "establishing contact with the child-to-be" (Svenaeus, 2017, p. 113). When we introduce imaging technologies into the relation of the parent with the yet-to-be-born child, things may start to change. It is as if the visibility afforded by ultrasound has dissipated something 'other' that only the mother can feel. As the mother lies in bed feeling for the presence of her child, technology has already touched her pregnant being. The child who remains 'in there' becomes 'distant.' We may wonder if fetal imaging technology does justice to the in-visibility of the child. We may ask if that visibility of such imaging only offers a specter to what is ultimately invisible. We may problematize an ultrasound image that views an organ rather than a child (and that near completely absences its mother). We need to continually question how our conversations and technologies may affect our responsibility to the child who has yet to be born.

NASCENT EXPECTATIONS AND HOPE

Before the discovery of being pregnant, there may already have emerged expectations: the bodily changes of pregnancy, having a child, and becoming parent. We expect that these expectations may be tentative, ambiguous, or even unrealized. Just consider the titles of popular commercial pregnancy books, written to appeal to expectant parents' sensibilities: *What to Expect When You're Expecting; Expecting Better: Why the Conventional Pregnancy Wisdom Is Wrong—and What You Really Need to Know*; and, *The Expectant Father: The Ultimate Guide for Dads-to-Be*. What is the texture and meaning of such expectations? How are expectations given in anticipation or hope? And when the unexpected occurs, what might the unexpected tell us of the ethics of expectations?

> It was actually the nurse or the receptionist that called, and she's like, "No I don't see anything on your file. The doctor just says you need to come in tomorrow morning." So, then I am thinking, there's no way a doctor is going to call you and tell you to come in unless something is wrong. My boyfriend said, "Don't worry. I'm sure it's nothing." But I lay awake in bed, tossing and turning, unable to get to sleep.
>
> The next day the doctor took us into his office and laid it out. He said, "It's gastroschisis" and "that there is a hole inside of the belly and his bowels are kind of hanging out." I don't remember all of the details. It was all so scary. But I wasn't really focusing on what the doctor was saying. I was focusing more on what my boyfriend was feeling. You could see the look on his face. I could tell, like, he wasn't fully sure about this, and perhaps about me.

Pregnancies come from all manner of relationships, as varied as the parents that conceive them. Not all mothers will have a partner that will see their pregnancy through. And indeed, some women will choose to become pregnant without a partner, to have a child of their very own, or by themselves. Shared between pregnancies is a sense of forthcoming, meaning the future is about to happen. While such a sense of possibility may reside in the fore and back of a parent's mind, the possibility exists for the unexpected to be discovered. Newfound information or knowledge may interrupt expectations: disturbing, displacing or otherwise disrupting a parent's sense—expectations and hope—of their child-to-be, themselves as a parent, as well as the family they could become.

Expectation is from the Latin *expectare*, expressing "an awaiting" or "await, look out for" (OED). What we expected is not simply information but rather a happening—something coming to pass. After all, we make plans in expectation, we get ready in expectation, we wait in expectation. And we also have expectations of what we expect, united to the possibility for unfilled expectations. We know that in a so-called uncomplicated pregnancy a parent may watchfully wait for their body to change, to feel the baby's first movements, and eventually, for the baby to be born. But when the unexpected comes to pass, the entire texture of expectations may be disrupted. It is not that expectations are simply inadequately filled; rather, that which comes to pass is of a different character compared to the expected.

> The diagnosis was Edward syndrome. I remember feeling the room was closing in on me. That I just could not catch my breath as my eyes teared-up. I was worried. Would he be able to walk? Would he be able to talk? Could he go to school? I was worried he would not be able to connect with his siblings and extended family. I thought about our other children, how would I be able to care for them? I remember just feeling so alone like we were the only people to ever expect to have a child with medical problems. Are we going to have to sell our house? I look back now at those pictures of me from those first days. I can tell I had been crying for hours.

A parent may respond to what is unexpected in a myriad of ways—sadness, worry, fear, confusion, or denial—as that which is un-expected undoes existing expectations. And we appreciate that as a parent comes to grips with new expectations additional worries may surface: their other children, their partner, their finances, and so forth (Gaucher & Payot, 2011; Gaucher et al., 2016). The parents may find themselves in a situation of grieving both their experience of pregnancy and their expectations of parenthood (Payot et al., 2007). And certainly, we need to appreciate how medical information given with more or less clarity or compassion may shape parents' emotive responses.

What is unique, however, to the natality of pregnancy is that pregnancy in itself gives birth to possibilities: the possibility to be a mother, the possibility to be a father, the possibility to raise a family together, and all manner of other imaginable possibilities. So much so we could say that the phenomenality of pregnancy expresses natalities of possibilities. The possibilities are expressed in what is expected and what cannot be expected, because such expectations may not have existed prior to conception of a pregnancy.

> I recall getting the results of my first-trimester screening. The doctor sat us down and said, "The nuchal translucency, a space in the back of the baby's neck, is enlarged. They also could not see a nasal bone. It could mean a genetic condition like Down syndrome or another trisomy. And even if the genes are fine, it could be a heart problem." My partner asked, "What are the chances that the baby is just fine?" "Five or maybe ten percent," she answered. What followed was weeks in limbo, waiting for test results and follow-up scans. I was not ready to talk about ending the pregnancy. But I was also not ready to find out the gender. I wasn't sure if I could keep it.

Related to expectations is hope. To live with hope is not merely to live with wishes or desires (for things 'hoped' for). From late, Old English, *hopa* means "confidence in the future" (OED). Hope expresses living with a sense of possibility. We recognize messages of hope in stories of striving to survive, persevering over hardships, or otherwise living toward the future. Neonatal-perinatal medicine researchers report that NICU parents want to hear messages of hope (Boss et al., 2008; Young et al., 2013), and in turn that health professionals have certain ethical responsibilities to support expectant parents in their hopes (Boss et al., 2008; Miquel-Verges et al., 2009; Payot et al., 2007). To be clear, this does not mean that healthcare professionals should hide the reality of medical diagnoses to create false hope. Most parents say they want to be told the truth, even if it is difficult or even devastating (Kavanaugh et al., 2005; Roscigno et al., 2012). Rather, supporting hope means supporting a sense of a future, acknowledging the possibility for survival even if chances are low, and certainly recognizing the possibility for a meaningful life even if that life is brief and supported in palliation. In other words, supporting hope does not mean being disingenuous but rather that healthcare professionals "acknowledge and even anticipate the possibility of good outcomes as well as bad ones" (Lantos, 2018, p. 9). We need to realize the consequential nature of medical diagnostic and prognostic knowledge. We do not merely find information about a child but instead we create and close expectations for the family, which carries the possibility of joy or sadness.

Implicit to the practice of neonatal-perinatal medicine is a culture of 'risk,' as healthcare professionals recognize that conception, pregnancy, birth, and so forth all have the possibility of complications.

> We were finding out more information than we really cared to know. On the screening ultrasound was the report of absent nasal bone. So, with that, we were told that our risk of Down syndrome had increased from 1/560 to 1/12. I do not remember what else our doctor said. I was too much in shock. My head was a flurry of thoughts and feelings. 1/12 was a real risk. We did not want an amniocentesis, as we did not want to risk the pregnancy. After much

talk, we decided to pursue the Harmony Testing. During those weeks that followed, I tried to just keep to work and other routines. And then the phone call came, "There is a 98% risk he has Down Syndrome." I held the tears in, trying to finish work. I felt so disconnected from our baby. It was like from that one phone call I had this being growing inside me who I no longer knew.

The etymology of risk—from *riscare*, meaning "run into danger" (OED)—reminds us a moral value is placed on any outcome we conceive of as a risk. We do not talk about the risks of winning the lottery, the risks of receiving a gift, or even quite simply the risks of it being a sunny day outside. Instead, we use the language of risks for undesirable outcomes. In neonatal-perinatal medicine, we need to ask what happens when even before a baby is born we talk about a child's life as a risk. What value are we placing on the life of a child with a genetic condition when his or her parents are counseled on the risk of him or her being born? How do we affect the manner in which a child is expected and hope for when we talk about him or her as high risk? What value are we placing on a life with a disability when we disclose disabilities as risks? Is it appropriate to place life possibilities for a child and family in such a fashion?

From a phenomenological perspective, we may recognize that words reverberate with meaning as we use them to reveal and reflect on the lived throughness of the ordinary and extraordinary moments of our lives (van Manen, 2014a). And yet, we also recognize that when we use words in conversations we do not necessarily deliberately reflect on their full meaning. As physicians, nurses, or other healthcare professionals, we may give little thought and attention to how we use words such as 'risk,' as they have become part of the everyday discourse of medicine, nursing, and allied health professions. From a bio-ethical perspective, if we pause, we may realize that reducing a child to a risk is profoundly unethical. The unexpected may disrupt existing expectations, however fragile they may be. A parent may experience a sense of detachment of the child that they thought they knew.

On the one hand, we can blame medical technologies whereby we recognize they inherently make the calculation of risk possible; it is as if the technologies themselves bear moral responsibility (Verbeek, 2011). On the other hand, such technologies and how they are integrated into clinical practice reflect an ethics of technology application. For example, ultrasound makes it possible to suspect a genetic diagnosis. 'Soft markers' such as nonossified nasal bone, linear arrangement of the tricuspid and mitral valves within the heart, thickened nuchal skin fold, relatively short humerus or femur compared to head size, echogenic intracardiac focus, fetal hydronephrosis, and so forth are all associated with aneuploidy (abnormal chromosome number) (Norton, 2013). Similarly, biochemical measures from the expectant mother's blood or urine, such as free beta human chorionic gonadotropin (FbhCG) and pregnancy-associated plasma protein A (PAPP-A), also have clear associations with genetic conditions (Tørring, 2016). When we combine maternal age, ultrasound findings, and biochemical values, further refinement of probability for aneuploidy can be calculated to determine risk (Alldred et al., 2017).

From the determination of risk, either chorionic villus sampling or amniocentesis may be offered for more definitive diagnostic testing, recognizing such tests have risks in themselves: precipitation of labor, needle injury to the fetus, and infection transmission. Alternatively, cell-free DNA detection methods for non-invasive prenatal testing (NIPT) are available such that genetic conditions may be identified with a high degree of certainty through analysis of tiny amounts of fetal DNA circulating in the expectant mother's blood (such as the above mentioned Harmony Prenatal Test) without the risks of pregnancy loss (Dondorp et al. 2016; Morain et al. 2013). To be clear, lack of invasiveness describes physiologic invasiveness relative to chorionic villus sampling or amniocentesis.

What is challenging from a bio-ethical perspective is that prenatal testing may disturb nascent expectations such that the yet-to-be-born may or may not be experienced as a child to be supported in his or her living. Since the advent and popularization of such screening tests for Down syndrome and other genetic conditions, the faces of such children are becoming rare as more and more parents choose to terminate their pregnancy on receiving such diagnoses. It is not unusual, and often routine, in some centers that families are given termination dates at the same time as the test results. It is as if a test result necessitates action—that is, pregnancy termination. After all, the test exists to spare future human beings unnecessary suffering (Milunsky & Milunsky, 2016). Although doctors, nurses, and other health professionals may tell the families that they have time to decide, and ultimately that they do not have to choose termination, this choice is made conscious and real with the test result.

Perhaps we use the language of 'terminating or ending a pregnancy' precisely because what these tests give is a particular perception of a child that replaces or diminishes what was ambiguous yet nonetheless present of the child that we struggle to name, feel, or otherwise contact in closeness. For the mother who is 'disconnected from her child,' we need to understand that what was given as a 'risk' was imbued with negative value that may have in some way or form disconnected her from her child. And of course, we can use phrases such as 'therapeutic termination' to confer on ending a pregnancy a sense of 'treatment or healing' as we recognize that a medical justification for abortion may exist in cases of fetal unviability, projected effects on quality of life, or risk to the mother's health. Important to debates around the morality of pregnancy intervention is the question of how decision makers perceive the fetus. What might be a parent's sense of personhood of the unborn child? And of course, how do different technologies, and the culture of neonatal-perinatal medicine, present or portray the child?

While we cannot address societal questions on the life value of individuals with genetic conditions, a phenomenological bio-ethics can nonetheless offer insights and raise additional questions. Pragmatically, healthcare professionals need to be aware of the implications of the language they use and how they talk to expectant parents. Even using a term such as 'chance' compared to 'risk' confers a different sense of value to the life of a child. We need to be aware that as healthcare professionals we instill moral value on the life of a child with a prenatal medical diagnosis simply by disclosing the diagnosis at the same time as the option of termination.

Regardless of the medical technology employed, we need to realize that the medical information itself gained from these technologies and how we communicate it to parents may dramatically affect the natality of pregnancy—the expectant, hopeful living with possibilities for a child to have his or her own meaningful life. We realize that what may seem as banal information for healthcare professionals may be highly consequential for families. And yet, as healthcare professionals, we also recognize that such medical information does not belong to 'us.' And so, while healthcare professionals cannot ultimately choose to keep such information private, we should be mindful of how we disclose it. Fundamentally, we need to appreciate, then, that all prenatal conversations on the yet-to-be-born child are highly ethical, as they affect the knowingness and actions on the child the parents may or may not expect.

MEANINGFUL OUTCOMES

Antenatal consultation is a core activity of specialists in neonatal-perinatal medicine. The intent is for the practitioner and parent to meet before the birth to contribute to the care of the mother and also the newborn. Consults may serve a multitude of purposes such as obtaining informed consent, relieving parental anxiety, elaborating treatment plans, and supporting parenting practices, depending on the various situational and contextual factors underlying the consult. The bioethical literature has typically focused on those antenatal consultations directed at decision making for anticipated preterm delivery approaching the 'limit of viability' at 22 to 25 weeks of gestation.

> A physician came to talk to me before Molly was born. I do not remember everything that he said except that we did talk about her chances. I was told if we resuscitated, we had a 17% chance of death or major disability. We talked about cerebral palsy, blindness, deafness, and other problems that could affect her quality of life. Of course, if we did not resuscitate, she had a 0% chance of anything at all. I do not remember all of it except thinking that she did have a chance. Nothing was for certain.

Molly's predicament is an absolutely profoundly significant matter for conversation, the chances for survival and disability. From a rational decision making perspective, we expect to inform a momentous choice, to attempt or not attempt resuscitation. And this conversation is profoundly ethically laden. Physician and parent are engaging in deliberation to discuss this very child, who while not yet born, is coming into being as a child. Resuscitation might or might not lead to this child having a life; palliation certainly will lead to death. What is the meaningfulness of a child's chances? How is disability understood when joined to death as an outcome? What is the significance of treating the life of a child as an outcome?

We may begin by considering what constitutes an outcome—that which results from an action, event, or other happening. Surely there are some outcomes that are immediately given. If the physicians attempt to resuscitate, the child will receive the interventions of intensive care immediately after birth, some of which are possibly painful or the cause of suffering, with the foreseeable need to continue such interventions for days, weeks, or months in a critical care unit. And yet, with resuscitation there is the outcome of life; the child may live, however long such survival may be. Then there are other outcomes for which we often justify intensive care, a future life (beyond the newborn intensive care) that may or may not in part be composed or otherwise affected by health and/or developmental challenges. All are outcomes! And we recognize they are determined with 'chance' yet also with exactness and precision, 'a 17% chance' that this yet-to-be born child might survive and live a life without what is described as 'major' disability. However, if this is what the physician aims to convey, is he speaking truthfully? Are these 'this' infant's chances?

It is doubtful that this '17% chance' has been estimated from knowledge of all of the predictive variables or causalities for prognostication: whether antenatal steroids were provided, whether there are signs of fetal compromise before birth, whether the fetus appears appropriately grown, whether resuscitation will be conducted by an experienced team, and so forth. Moreover, it is unclear whether the physician considered with what measure of exactness he determines the variables informing his prognostication. For example, measures of maturity, such as gestational age, are fraught with error. First-trimester ultrasound gestational age estimates incorporate a standard deviation of 4 to 7 days, with the standard deviation becoming wider as pregnancy progresses (Chervenak et al., 1998; Saltvedt et al., 2004). So, 17% is only an estimate, anchored in confidence intervals or some other statistical means used to convey probability. And yet, given as a number, 17% is deceivingly precise.

Perhaps most importantly, we need to acknowledge that 17% is not independent of the decision maker. This child's chances are not the same as the one in six (17%) chance of rolling a given number when a six-sided dice is rolled. This is because this child's chances are ultimately dependent on the family's values, beliefs, and intentions. The chances will change depending on what the parents are willing to accept as acceptable actions and outcomes: Are these the kind of parents that believe in life at all costs and are therefore inclined to pursue all manner of medical intervention regardless of invasiveness? Would they potentially consider limiting or discontinuing medical interventions if injury to the baby's brain or other organs was found?

Then again, we can appreciate the efforts taken on the part of the physician. Estimating chances for survival is part of the culture of neonatal-perinatal medicine, leading some to even devise prognostication calculators (Tyson et al., 2008), and professional bodies to endorse the appropriateness of communicating such risks in antenatal conversations (Cummings et al., 2015; Lemyre et al., 2017). We can appreciate the physician's desire to cite, fathom, or appeal to some valid estimate for survival that the parents want to know.

But decision making and established policies for resuscitation and medical intervention do not just reflect outcomes—they also shape them (Janvier & Lantos, 2016; Janvier et al., 2017). And we also need to acknowledge that the use of objectifying formulations of disability such as cerebral palsy, cognitive delay, or neurosensory impairments has a normative bias with none of these diagnoses directly linking to quality of life. The medical model directs us to think about disability as a "tragedy or problem localized in an individual body or mind" such that the meaning of disability is narrowed to a deficiency of the body requiring a solution (Beaudry, 2016, pp. 210–211). It also fails to disclose understanding that individuals navigate and exist in the world in a manifold of ways with disabilities. It is challenging to capture in language this complexity of meaning for the outcomes of intervention.

But let us return to the mother's words: "I do not remember all of it except thinking that she did have a chance." Her words reveal the phenomenon of chance. Chance is actually more exact than 17%. It is absolute in its possibility, transcending numerical formulation of probability and uncertainty. The child unequivocally has a chance if the parent chooses intervention.

> We had tried so hard to become pregnant, so when my water broke at 19 weeks, and I was put on bedrest, well it was more than my water that felt broken. Still we had to try. Initially, the NICU doctors did not seem to want to come. Apparently 19 weeks was just too early if she was born that she could survive. So, with each week that passed, each week that I did not deliver, I began to get some hope back. A future seemed possible for my child. But when I was 23 weeks, I was told the chances were still not good. Survival was unlikely; a chance of less than 10%. And then there was the chance of her being blind, deaf, unable to walk, unable to talk, or her life otherwise permanently affected by a premature birth. It was too much.

Perception studies tell us there is a difference in felt meaning to a 10% chance of survival to a 90% chance of death (Haward et al., 2008). And we know that comprehension of statistical information is difficult even for educated people. For example, people may understand 4 out of 10 differently from 40%, depending on the risk it expresses (Malenka et al., 1993; Mazur & Hickman, 1991). But beyond the challenges of understanding statistical information, and the textures of a good or poor chance, is the meaningfulness of outcomes.

Survival is laden with meaning. With the birth of life of the child comes the birth of the mother and/or father. Birth as genesis is the future as open, possible, and imagined. We could say that life does not give itself at once but instead offers the vital growing of more life. While death itself is finitude for the dead, an ending for what had begun, for the still living, death leaves a trace or a shadow that in its absent presence never actually dies. In this way, while life is a fleeting presence, death is eternal even in its absence.

Of an entirely different texture is disability, the possibility of a life marked by a difference that perhaps had never been conceived. Surely disability is not necessarily a life of suffering. Many children with disabilities enjoy good quality of life. And varied disabilities may differ in their meaning for a particular child's life. Disability is not a proxy measure for quality of life. Quality of life conceptually lacks clarity and objectivity such that discussions may progress to disagreements in principle (i.e. what an individual perceives as consisting of good quality of life). It is not necessarily a life without joy and meaning. And yet, we may give this other imagined life as something quite other to the life imagined in conception or felt in embodiment.

Now, death or disability is a commonly reported composite in outcome research for reasons of statistical power and statistical significance. And not surprisingly, such composite outcomes from the literature have been imported into antenatal consultation and other counseling activities in neonatal-perinatal medicine. Disability, however, has a very different meaning from death. It is not simply that to juxtapose disability to death presents a life with disabilities as an adverse outcome in itself (Janvier & Farlow, 2015; Janvier et al., 2016). Juxtaposing death and disability ignores the phenomenological complexity of giving the possibility of very different meanings in their outcomes, and vice versa.

Up until now, we have glossed over an ambiguity that is not disclosed by these parents' words: What child is 'this' that has a chance? Is it the imagined child of conception? Is it the embodied child of pregnancy? Has the consult changed the child to be? If the child is born and intervention sought, then his or her life will be affected by the absolute reality of medical interventions of intensive care. Perhaps, following Travis Rieder, we can say that 'resuscitation' is the incorrect term for the medical team to use. The physicians, nurses, and other health professionals are supporting the 'creation' of a life because they cannot completely restore the child as his or her life was imagined to be (2017, p. 5). They are supporting a new natality, the possibility for a child who survives to live a life different from expected. Is it all too much for 'this' child?

> After our consultation, we were offered a tour of the NICU. All the statistics of before … well, they just did not seem so important. I was not concerned about whether we could get him to survive anymore, whether he might have some problems in the long-term. Instead, it was all about, "Do I want this for my child?"

Although we may spend a considerable amount of time talking about outcomes such as death or disability, another outcome which we can be more or less certain about is the reality of intensive care: being in hospital for months, potentially connected to breathing machines, receiving nutrition by intravenous drip or feeding tubes, and so forth. Clearly, these are outcomes in themselves and certainly may inform decision-making discussions. Yet, these conversations about what intensive care might be like for a child are strangely absent from the medical literature. Instead, the antenatal consultation literature focuses on informational needs around such issues as: survival, development, parenting practices, and length of hospitalization (Alderson et al., 2006; Blanco et al., 2005; Haward et al., 2017; Studer & Marc-Aurele, 2016; von Hauff et al., 2016; Yee & Sauve, 2007). When medical care for a child is conceptualized as a part of antenatal consultation, it is reduced to needed medical interventions or expected complications. While such topics may all be appropriate for a particular antenatal consultation, do we not need to acknowledge questions of pain, discomfort, and suffering and the contrary comfort, consolation, and wellbeing as important outcomes to consider?

> I was in pain. I was bleeding. And they're talking to us about statistics, but none of it was getting through to my mind. All I could say was, if he comes out and he's not looking good … let him go. But if he comes out looking as though that he's up for all of this, then please save him. Whatever happens, I just don't want him to suffer.

The reality of perinatal conversations is that conversations may be curtailed by the realities of impending delivery, maternal illness, or so forth. Yet, ultimately, from a bio-ethics perspective, that does not mean that we do not need to reflect on how we talk about the possible lives of children who may require medical care. Before a child is even born, should we as health professionals be facilitating the reception of a child whereby their life is reduced to possible outcomes? Or should we use language that reflects a possible life, which might or might not involve certain outcomes? Clearly this distinction is subtle yet an ethical one.

> And it was then and there I realized. We cannot really know what is going to happen. We don't know when I am going to deliver, and we don't know how it will go for my baby. All we know is that at some point we will probably have to face some decisions.

From such a realization, perhaps we need to recognize that the notion of antenatal consultation is somewhat antiquated. Instead, we need to focus on perinatal consults that are personalized (Gaucher et al., 2016; Haward et al., 2017), improvised (Janvier et al., 2014), and conversational (von Hauff et al., 2016). Parents need more than information from antenatal consultation (Gaucher & Payot, 2011), and there is much that healthcare professionals can offer if nothing less than supporting parents as they face the possibility of their child needing medical care.

For the field of neonatal-perinatal medicine, outcomes for clinical conversations have traditionally been chosen based on reading of the literature. The problem for neonatal-perinatal medicine is that various outcomes (such as death and disability) have been "fixed into practice" without ever having been evaluated by families and other non-healthcare provider stakeholders (Janvier et al., 2016, p. 572). In other words, these supposed shared decision making conversations may open from a perspective that is fundamentally not informed by a particular family's perspective and, therefore, reduces consideration of what is in the 'best interests' of the child to a consequential appraisal of medical outcomes.

On the one hand, when we talk about chances as the result of decision making, we realize that we are talking about 'what' we are 'giving' a child when we give it a chance: life, death, disability, and so forth. On the other hand, we also need to recognize that what we are 'giving' a child from decision making also ensconces what a child may live through in the pursuit of an outcome: specific medical interventions, life in the context of a NICU, and so forth. In a sense, all of these outcomes are consequences, sequels, or effects. Yet, at the same time, the phenomenological experiential meanings differ. Finally, we need to acknowledge the awesome finality but also significant potential for error in prognoses underlying the decisions facing families and health professionals in neonatal intensive care (Duff & Campbell, 1973). The weight of such decisions affects the child, the family, and the community whether community is defined narrowly within the confines of the NICU or broadly as society at large.

SECTION II

In the Cradle of the Newborn Intensive Care Unit

NEWBORN ENCOUNTERS

The first time I saw and held my child I felt taken aback. It was so much more than I expected. Here was this little child who appeared almost like a little stranger. And yet, he was also instantly recognizable. I just stared at him. I looked at his fingers, his hands, his hair, his eyes, all of his little features, all for the first time. I was overwhelmed. There were so many things I wanted to see. Still, the more I looked, the more I found my gaze constantly returning to his face. It was not that he was looking at me, but I could tell that he felt me. Looking at his expression, he seemed to settle into my arms. He was mine. I was holding him in my gaze and he was holding me, and I did not want to let him go.

The experience of the newborn child may be like an aporia—familiar yet strange, recognizable yet new, anticipated yet unforeseen—more than the meeting of sense and sensibility. Language falls short to describe the ethical nature of this meeting with the newborn. When the mother speaks of being stirred by his face, it is not just the physiognomy of the child's face. Nor is it the so many little things of small hands, delicate hair, peeping eyes, or other fine features. The mother is held in an encounter of contact. The word 'contact' derives from *contingere*, meaning "to touch" in togetherness (OED). To face an other, says Lingis (1994), is to touch with the eyes, and the vulnerability of the other is felt in our eyes.

In a moment like this, the meeting between parent and child is not just a meeting but a true ethical encounter: encountering the child's self in his or her singularity. How can the mother be anything but taken aback when faced with the face of her newly born child? Levinas (1969/1961) helps us understand the ethical command of the face. The face in its irruptive expression "calls me into question" (p. 83) and also makes a demand on us. As such, the experience of the other, the face, becomes the condition for the possibility of ethics. The responsivity of the parent is called into question, held by the child, the other, beyond expectation.

Responsibility is the parent's ability to experience a response to his or her child's call: a parental ethics.

Another parent describes,

> I remember the first time that I held my child's fingers, or rather that he grasped my index finger and held it. The nurse had put him in my arms. He seemed so fragile. I just stared at him. As his head turned, his eyes seemed to be looking for something to hold onto. And his arms were jittery fingers groping at the air until they found and grasped my finger. I was struck by his face. I looked at him, somewhat confusedly stunned. I felt taken aback. I was holding this baby in my arms who was my child. I cradled him so carefully, keeping everything tucked in and safe—the feeding tube and monitoring wires, just so. Although I held him in very close, it felt like he was really holding me, so firm was the grip of his little fingers. As he moved his arms, I moved my hand within his reach, allowing his fingers to close around mine. I did not even think about these subtle gestures until they had already happened. As my child was holding me with his fingers and his gaze, I experienced a powerful and overwhelming sensation: This little baby was making me a father.

This is a father's first time to hold his child, born into the medico-technological world of the NICU. Indeed, if it was not for the NICU, the child might not be alive. But, for the father, this technological reality is simply a backdrop to his encounter with his child. In touching each other with their eyes and hands, in the crossing of gazes and caresses, the father and child find each other, making contact.

The grip of the child's finger announces his presence as an other to the father. The father feels the child's grip physically and feels this grip pathically, relationally and existentially as his own fatherhood. Buytendijk (1970) has suggested that touch is expressive of a direct kind of intimacy, as it establishes in a 'feeling' way a close relation. In cradling his child, the father experiences both the child's touching and his own touch. It is a double aspect of responsivity and sensitivity, as to touch allows for an experience of grasping understanding and opens the parent to a moment of being moved by the otherness of the child. It seems that in the contact of seeing and being seen, the touching of hands, the father encounters his own sense of fatherhood.

Emmanuel Levinas writes of the fecundity of the father-child relation:

> I do not have my child; I am my child. Paternity is a relation with a stranger who while being Other ... is me, a relation of the I with a self which yet is not me.
>
> *(Levinas 1969/1961, p. 277)*

It is as if paternity introduces a sense of otherness already within the parent. This otherness is not just strangeness but rather an evoking appeal belonging to the enigma of the child's being, an otherness that cannot be reduced to 'me,' the self of the parent. In the words of Levinas (1969/1961), the self of the "I is not swept away, since the son is not me; and yet I am my son" (p. 277). And as the child is not 'me,' yet is of the parent, the child is also the future, the infinite, the transcendent: "the fecundity of the I is its very transcendence" (p. 277). While we may stumble on such a philosophical formulation, we ought to pause to reflect on the fullness of meaning pointed at by Levinas. The relation of parent to child is not simply unique but necessarily unique, expressing a responsibility for a child of me yet other to me.

Husserl writes, "Among the human beings in my vicinity my child is the 'nearest' to me and this is equal to an irrationality of the absolute ought" (unpublished manuscripts, in Ferrarello, 2016, p. 214). What Husserl names as "irrationality" is not foolishness or absurdity but, rather, that which transcends logical thought experienced in its exemplary moral self-determinacy.

But it would perhaps be a mistake to limit paternity or maternity to genetic factuality. A mother responds to the crying call of her newly adopted child.

> I could not look at her as if she was just someone else's child. Although she was not from my womb, although she was not of me, I saw her and she saw me. She looked right at me. I could not placate myself that this was someone else's child and therefore that I could shed my duty to pick her up, to respond to her crying. I felt her. In hearing her cry, I was already responding. Perhaps, I did not have to pick her up. But, I did have to hear her cry. I could not leave her bedside even if I could not touch her.

Here, ethical responsibility arises not from the in-ability to sooth the child but rather from the child's demand on the mother. It is the experience of encountering the crying face of her newborn. What the mother responds to is not just the countenance of the child. The child's look and cry become a raw experience of proximity. She sees the child with a "listening eye" (Levinas, 1981/1974, p. 38). And it is as if the sound of the child's cry "overflows so that form can no longer contain its content" (Levinas, 1989/1949, p. 147). A single sensory faculty seems insufficient to perceive the fullness of the call of the child. It would be incorrect to take ethical responsivity superficially in the physiognomy of the child's physical features or in the shrill timbre of the child's actual vocalizations. As Jean-Luc Marion (2000) says, "To receive the face implies not so much to see it as to undergo the impact or feel the shock of its arrival" (p. 226). This is the ethical moment as evoked by the parental encounter with what appears yet remains absent from the visibility of the newborn–child face.

And yet for others there is a different texture to contact,

> The moment he was born, they rushed him off to the corner of the room. He was ashen. I watched as they put the breathing tube in, pushed on his chest to try to get his heart going, and finally had to resort to putting an intravenous

line into his abdomen. I never heard him cry, yet I felt sick in watching. When I was able to get in there, to stand beside his bedside, to look at him, I just wanted them to stop. If there had been a spirit of life in that body, surely it was gone now. He looked bruised and mottled. And his skin felt cold. Although they wanted to transfer him to another hospital for more treatment, I could not help but hope that his body would just die. I could not stand to look at his face. He was so damaged.

We may feel that we understand the pain and hurt expressed by the father in response to the encounter with his newborn child. But our understanding has to probe deeper. It has to gain a measure of the experiential substrate, where ethical responses acquire their meaning. In the words of the father, we sense that beyond the pain stir emotions and responsivities that are preverbal, almost impossible to put into language. We see how the father already expresses an ethical response in his hope that his son's "body would just die." But is it correct to say this? For us to say, "his son's body" already presumes a possible mis-interpretation of the father's felt responsivity. Did the father already experience the newborn as his son? Or is even this assumption possible in such a moment of chaotic mental distress?

I could tell something was wrong after he was born. It was like the nurses and doctors did not want me to see him. They were gathered around, talking about him, and as I walked towards them, it felt like they were holding him back from me. When I saw his face, I understood: he was malformed. I step-ped back. I did not want to hold him. Even in seeing him, I felt uncomfortable and uneasy, as if I needed to shake something off my skin. I could not help but look at him, but I also just wanted them to take him away. To put him somewhere where I did not have to see his face. Looking at that face, I just felt like he should not be there. I felt like someone else's child had been placed here in place of my child.

Responsibility may be experienced as pain, too much for a parent to bear. In this account, the father's reaction is visceral, as he has already been touched by what he does not want to touch. It is a child from him yet also so different from him: a face that he can neither look at nor look away from.

And yet another encounter:

For some reason, I was still expecting to see my child swaddled in a blanket or lying tucked into the so-called fetal position. I was expecting her to lie there peacefully even if she was small or sick. But when they brought me to her, it was hard to see her. I had to look in at her from the incubator portholes. There were wires and tubes obstructing the view of her skin and hands. I was looking for something human in there. I was looking for her face.

For the baby requiring medical care, the relation of parent and child has the capacity to be complicated by illness. And as an abundance of medical technologies are also introduced, some highly sophisticated and others seemingly ordinary, the parent may struggle to make contact. The child is more than buried beneath, behind, or within this equipment. The parent enters an unexpected ethical place where materiality obstructs the parent's first touching encounter with the child. And in the absent visibility of the child's face, the parent is touched by the tangle of technologies.

We need to wonder about the ethics of what happens when perception and touch become mediated by illness and technology. In the NICU, the child is rarely experientially given to the parent free from such consequences. Rather, the child is absorbed into the technical environment: contained in isolette or bassinette, connected by lines to machines, displayed on monitors. How, then, does a parent perceive his or her child? How does a parent experience a child that seems wired into this medical-technological world? How is the child of technics given in experience? What can we learn about the parent's experience of response? How is technology seen or perhaps not seen? What are the ethics of the cradle of the NICU?

> Before he was born, we decided that we were just going to focus on his comfort. At the time, the heart surgery they described sounded like too much for a child to bear, particularly given the need for additional major surgeries with time, and the possibility of complications. Yet, when we held him, when we saw his face, we realized that we could not live with that decision. We could not let him go.

There exists an ethical imperative to support or engage the newborn, "whose mere breathing uncontradictably addresses an ought to the world around, namely, to take care of him" (Jonas 1984, p. 131). It is not just that the ethics of neonatal-perinatal medicine may be complicated by technology and illness but rather there is a natality to these ethics, meaning an inceptual unfolding of ethics from the conception, through the pregnancy, to the birth of the child who comes to reside in the NICU. Conversations before and after birth need to anticipate and respond to the ethics of conception and pregnancy, and ultimately how the presence of the newborn may profoundly change the perinatal decision making reality of all partners and players in the healthcare setting.

Care may consist in medical intervention or comfort with decision making subject to contingencies, uncertainties, and unsuspected sensibilities. There are so many personal accounts of new parents describing how when they first held their baby "everything changed." We recognize "the ethical standing of the embryo becomes more powerful and demanding as it develops beyond the stage of implantation and gastrulation and is in place in the uterus of a woman" (Svenaeus, 2017, p. 120). We may wonder, how does ethical moral standing grow when the child, born prematurely, is now outside of the womb? How does a child in the

NICU exercise a different demand from the child who remains en-wombed? For many parents, discussing the withholding of medical treatments is a fundamentally different relational experience when a baby has been held and touched compared to prior to birth. Healthcare professionals, too, may be stirred by changed relational responsibilities to infants before and after birth. In other words, the birth of the newborn may change everything (including the sense of autonomy). Conversations with families should not simply support and respect the relational autonomy of parent decision makers. Ethical conversations need to anticipate and consider the increased relational complexities of parent, healthcare professional, and, pointedly, the newborn child.

TECHNICS OF TOUCH

In a few days, it is going to be 3 weeks of him being here, stuck on a breathing machine. Still, even up to now, though I am here every day, I don't feel like a mother. I sit by his isolette, look through it, open it, reach inside it every day just to be with him. But it does not seem to help. I don't know what to do. There is too much going on in there—tubes, wires, and needles, everything. Sometimes I just stand here and talk to him. I tell him that he is doing a good job, but I don't know if he can even hear my voice. At times, when I put my hand in there, he grasps it, holds onto it, and won't let go. He won't do that for my husband. He won't do that for my mom or my dad either. I am gathering that he knows it is me. But it is hard to say. There is just so much around. I mean, there is a nurse on this side and another nurse on that side and my husband behind me, talking to me. And I am just trying to be with him. I mean, I just wish they would all shut up, and let me just be his mother.

Sitting beside the isolette, doing all she can to have contact with her child, the mother finds herself unable to be a mother in the deeply felt sense of motherly belonging with child. Although the mother is bodily present, she still finds herself off to the side of the isolette reaching in—looking in at her child without facing the face of her child—unable to touch him with her look. It is true; the mother can open the isolette. She can reach in to touch her child. But this is not necessarily the touch of contact.

It seems like the mother needs to overcome the technological barrier of the isolette and its medical accouterments. She is aiming to do just that by trying to look at her child again and again, day after day: speaking to him and touching him. But in spite of all this, her trying does not produce the contact she really seeks. She seems neither able to be with her child nor to encounter her child in this deeper sense. She is rather reduced to looking at her child, who, in her words, is stuck in the isolette on the breathing machine.

We may understand the isolette as a structural thing that houses a child: an external plastic womb (van Manen, 2012b). It is designed to be a contained and controlled place of incubation, providing warmth, quiet and humidity until the child is ready for the outside world. As a medical device, the isolette is a high-tech plastic box that facilitates medical care. Routinely, such devices are used for infants less than 1500 grams, as well as, in some situations, for larger infants (Prescott & Hehman, 2017). The isolette holds in its inner world the body of the child and a host of medical things.

There are technological artifices directly connected to the child: intravenous lines, breathing tube, and monitoring wires. There are other devices placed for caring: syringes, suction catheter, and stethoscope. In this way, the phenomenality of the isolette is not just to enclose the child but also to unite the child with the technological: clinical monitoring, artificial nutrition, and various medical therapies. In a Heideggerian sense, the isolette not only gathers what it concretely contains; there are also 'things' that are immaterially gathered, as the technology seems to open the parent and others into a technical way of being with the child (Heidegger, 1971). The gathering is an existential assembling, a bringing together, of the child with the technomedical environment itself. For the prematurely born newborn, the technological sustains the child in a clinical place, affording the premature transition from womb-world to isolette-world.

The isolette invites the outsider to look in by virtue of its transparent fashioning. But when the mother reaches to touch her child, she encounters a barrier, the wall of the isolette. To enter the isolette, she must first unlatch the door. And even after opening the isolette, she still experiences an impediment. The isolette only permits the use of hands and forearms. No real hugging or holding. No true embrace. The isolette constrains and constricts, restrains and restricts. In addition, the isolette only permits a particular kind of touching. It does not support the touch of intimacy. The child cannot be held within her motherly arms. The child is held at arm's length.

Although the isolette may invite the parent to look into the isolette or handle the child in a particular reaching fashion, there remains a containment to this container. With the doors closed, the isolette dampens the audible urgency of the child's cry. With the cover laid over it, the child's very bodily being is hidden. To the onlooker of the closed and covered child, he or she may seem shelved away, stored securely in its placement of the room. In this way, the isolette ambiguously prevents nearness as the incubator becomes a sovereign place so that handling the child may seem disturbing to the child.

As a technological enclave, the isolette is not only an isolating thing, as its connectors and openings permit it to be connected to the technical paraphernalia of the NICU. For the parent, the isolette may come to represent the sights, smells and sounds of the NICU, as being with the child may become a familiar experience of monitoring screens, antiseptic aromas, and machinery sounds. There is a multistability to this device. It may be removed and replaced, resettled into various relations, but once enclosing a child, it becomes stable and constructive, constituting aspects of a larger system of relations (Ihde, 1990). This multistability is not the same as neutrality: "Within multistability there lie trajectories, not just any trajectory, but partially determined trajectories" (Ihde, 2002, p. 106). For the isolette, the trajectory is this system of relations, which may involve other objects (infusion pumps, ventilators, monitors) and people (nurses, respiratory therapists, doctors).

The child may not be easily removed from the isolette without the help of nurses or other medical staff. The operation of the isolette and its connections are the expertise of these professionals. The parent may then find him or herself at the mercy of others to do the simplest of tasks for the child: changing diapers, feeding, and so forth. Even touching the child may be felt as a supervised touch under the watchful eyes of others. We can see how the mother may experience herself as an outsider to the relationship of nurse and child. The experiential sensibility of maternity and paternity may become tenuous, fleeting and elusive as the intimacy of true contact remains out of reach.

In comparison to the isolette, technologies like the feeding tube seem rather easily passed over, taken for granted, as they more subtly weave into the tactility of technical flesh constituting the medical life world of a parent and child. Perhaps it is due to their plainness that we easily forget that they too entangle and disentangle the touch of a parent's encounter with his or her child.

The feeding tube is one of neonatal medicine's simplest of technologies. It is a flexible, silastic hollow tube usually inserted at the bedside by feeding it into the child's mouth or nostril until its end reaches its destination, the stomach. Secured in place by taping the exterior end to the angle of the mouth, the feeding tube becomes an extra body orifice, mechanically connecting stomach to world, offering a new way to touch the child. Bypassing the voluntary swallowing, the tube orifice is existentially different from the mouth orifice as food and medicines can be pushed slowly or quickly into the stomach by means of hand syringe or infusion pump. In neonatal care, this technology is used extensively as many infants are unable to effectively coordinate suck and swallow due to prematurity, illness, or some other issue.

> Yesterday was a bad day for Amy. Her feeds had been put up quite a bit. More than I would have liked because she spits up a lot, and you can see that she does not like eating. Her feeds are my pumped breast milk, all given through the tube, running in over two and a half hours. She gets nothing by mouth. And she has to eat every 3 hours so that she really only gets half hour breaks between feeds. So yesterday, again, she was spitting up pretty much all day. I was just covered in it. And she was crying a lot, in pain. Every time the feeds started again, she would start fussing and get really cranky because she had no room in there left. The nurses and doctors, they were not listening to me. I totally understand my child needs the feeds. Really, all I want is for her to be able to just handle the volume and grow. But forcing it through causes her pain. And the nurses and doctors, they just keep going, pushing the feeds

down into her. I finally said, "I am tired of this. You guys need to slow this
down. Just look at her!" They agreed and gave her a break.

As feeds are pushed in, both child and mother are in pain. The child seems
unable to handle the volume, expelling what is forced in. Although this feed may
be the milk of the mother, it does not appear to be experienced as from the
mother.

It may be tempting to consider the feeding tube on a spectrum with bottle
feeding, as both devices afford provision of milk by another person without the
intertwining bodily contact and communication between mother and child epito-
mized in the gesture of nursing:

> Never have I been in such close contact with another being's skin, arms, and
> mouth than during those early weeks of continuous holding and feeding. I
> made milk, smelled like milk, was sticky with this stuff that was me, but not
> me, which produced in me the need to give it away.
>
> *(Simms, 2008, p. 11)*

Still, the feeding tube does more than circumvent the bodily relation of infant
mouth to mother's breast. It opens the child to a new technological intentionality
whereby milk may be given, pushed into the child, in a technical way.

To provide food by a silastic utensil alters the natural breast and even the
pseudo-natural bottle experience of feeding: To touch the child with the intent of
nourishing becomes a gesture of contact with the feeding tube. The child cannot
seek, suck or turn away from the nipple to control whether or when to feed, how
much to feed, and at what pace to feed. The vocabulary of latching, pacing and
burping is no longer descriptive of the experience. Instead, everything is programmed, controlled and set, as the feeding pump buttons read: start, stop, volume,
and rate. Feeding becomes scheduled with time on and time off. If feeding is tol-
erated without issue, the child's being is reduced to a stomach: a passive receiver of
programmed nutrients and medicines. The routine of checking if the child is ready
to feed can be reduced to ensuring that the feeding tube is appropriately positioned
and secured. The child may not even need to be held, as the tube affords an
extended distance of feeding. In this way, the tube becomes a bodily orifice of a
cyborgian-child.

For the professional and perhaps the parent, this care at a distance has the
potential for disembodying and passing over a more originary, ethical kind of
caring touch that is encountered in a direct face-to-face encountering relation of a
caregiver and child (van Manen, 2002). Still, an ethical care of responsivity and
responsibility may transcend the mere provision of nutritive support.

> There are things that I notice and know about Amy because I sit with her all
> day, every day. I know what she reacts to and how she reacts to things. I
> know she needs to sit up when you are running the feed, that you can't rock

her; and that sometimes when she is upset you need to just hold her. When I have to leave, I find that I just can't leave because the nurses and the doctors do not see her as I do. To leave is to leave her alone in pain. There is nobody who really knows her; nobody to hold her like a mother can hold her. It is just that I get so tired and I need to sleep. So, on days like yesterday, I stay as late as I can. But even then, it never feels like long enough.

The mother sees and feels the child's pain with her motherly empathic body. When the child struggles with tolerating the feeds or the feeding tube, it may be difficult for the parent to watch his or her child struggle against the feeding tube insertion: thrashing, retching or literally grabbing to pull out the tube. Once inserted and secured, the child may throw or spit up the tube along with any undigested fed residual contained in the stomach.

The mother sees and worries. Yet, we cannot so easily accuse this technology of disrupting parental sensitivity and sensibility. Obviously, the child is supported by the technology of the tube for its critical medical needs. The mother is interlaced with a technoscientific world of seeing her child, of knowing her baby. Her maternal body seeks for a genuine contact with her baby's body. It should be possible to accommodate and harmonize these different needs. The tube need not cause a disembodied relation but rather could mediate a new form of a technically embodied relation, a relation where responsibility of contact may remain. As the feeding tube is woven into the relationship of parent, professional, and child, the professional may need to heed the mother's reminder to also "Just look at her!"

We may wonder further if there is more to this new feeding tube orifice, as it opens the child to a technical way of feeding, and it opens and broadens the nurturing relation to a host of others—physicians, nurses, respiratory therapists, and dieticians—who likely see the child differently than the parent. The tube can turn into an issue for quarrel as the healthcare professional oversees supplies, directs feeding, and gives the orders to increase or decrease the volume. Thus, the parental expertise of knowing the child maternally needs to be balanced with the medical expertise of knowing the child's physiological needs.

There are of course a multitude of technologies in the NICU mediating the manner in which a parent comes in contact with their child and that we could explore. The point is that these technologies (as most technologies) are not merely instrumental in their consequences: They carry ethical significance. Nor do the technologies uniformly affect the manner in which the child is present to us. Whether we are talking about sophisticated technologies, such as extracorporeal life support (ECLS), or the most routine of technologies, such as a cardiorespiratory monitor, they all have ethical significances (van Manen, 2015).

A distinction may be made between the ethical significance and the moral relevance of technologies. The moral relevance speaks to the way that technical devices open up (effect) new possibilities for human actions in the sociopolitical order of things. In medicine, the use of certain technologies is creating moral dilemmas. For example, the availability of the isolette and associated equipment affects the opportunity to provide medical care to infants born extremely premature at the threshold of viability. And as a moral dilemma, the parent and professional must then weigh the implications of these technical innovations, as the risk of morbidity may call into question the appropriateness of technological intervention.

The psychology and anthropology of touch and bonding signifies the importance of the subtleties and complexities of contact. But the experience of holding, touching, gazing and mutual sensing may be hindered or helped by the NICU technology. The newborn in the isolette may lack the opportunity for this primordial contact that is critical in the shaping of the deeply human. The feeding tube may be medically necessary and yet obstruct the sensation of pleasure, desire, and rhythm of the nurturing child and the mother.

Merleau-Ponty speaks about the interplay of touch and perception as humanizing crossovers.

> A human body is present when, between the see-er and the visible, between touching and touched, between one eye and the other, between hand and hand a kind of crossover occurs, when the spark of the sensing/sensible is lit, when the fire starts to burn that will not cease until some accident befalls the body, undoing what no accident would have sufficed to do.
>
> *(Merleau-Ponty, 1993, p. 125)*

I have tried to raise issues of the modalities and meanings of the crossing of touch and gazes in the technics of contact in the NICU. Stiegler (1998/1994) has suggested that technics, the technologies that keep us warm and fed and comfortable, are the ancient humanizing forces in the history of humanity and the development of human communities. In the NICU, the ontology of technics is the medical expression of these old and modern anthropological technologies. We need to be cognizant of how these technologies may put in question the humanizing meaning of contact between parent and child. Of course, it should be clear that the highly technologized environment of the NICU is not inherently problematic, as these technologies all have their important healing and lifesaving clinical uses. The incubator is critically important to provide an environment for supporting the child, just as the feeding tube may be necessary to provide nutrition for the child. So, as a neonatologist, I do not mean to question the utility of neonatal medical technologies but to show the possibility for technologies to have more than their intended effects. Lucas Introna writes on the value of making the so-called scripts of moral actions of technologies explicit: "we ought to 'open up' the complex black boxes of our technologically advanced society and 'read them out aloud' – in a language accessible to those who may potentially be enrolled in them" (2019, p. 28). As an example, isolettes could be fixed with labels of their effects on human contact in the same way that cigarette boxes bear health warnings.

SKIN-TO-SKIN TOGETHERNESS

The techno-medical environment of the NICU, and the medical condition of the child, may be overwhelming for the parent.

> It is really nice being able to finally hold her against my skin. To lay my hands on her back and feel her breathe rather than look at her through the incubator. Still, it is terrifying. I am constantly waiting, holding my breath, for an alarm to ring, or something to go wrong. It is hard knowing what to do and whether I am doing the right things. I have been expecting her, expecting being a mother, but I just don't feel prepared for any of this. I don't know how to hold my child even though I know I need to hold her.

For the parent, it is not simply the technomedical sophistication that is hard to deal with; it is 'my' in-ability to hold 'my child.'

> Kangaroo care, holding him skin-to-skin, is our time. It's so settling to feel his warm skin against mine. I get him nestled in, on my chest, and just lay my hands over him. His breathing steadies. He holds a breath, I hold a breath. I find myself sighing without even meaning to just as he exhales. It's like my body senses his and harmonizes with his. I do not normally even give it any thought. We just are together sensing each other's hearts beat. Sometimes I will read a book with my free hand. Other times, a friend will be there and we talk quietly. I don't really need to concentrate on him; I am feeling him as he is feeling me. To have him close – it just feels so good, so calm. Sometimes we just need to lie there together and let the day pass by.

Despite the technically specialized environment of the newborn medical nursery, the mother is able to just be with her son in almost (in)voluntary automaticity. It is as if all of the wires and tubes, the pumps and monitors, the nurses and doctors, fade into the background. The mother is in touch with and touches her child just as she is touched by him in a perceptual crossing of bodily presences. They lie attached in touch. They breathe in touch. They are touched by each other's touch. And from this attached being of touching and being touched emerges the pedagogical being of mother with child. It is a sensitive being whereby the boundaries of self and other are blurred. The interplay between touching and being touched reflects an elusive exchange of sense and sensibility. We may recognize this reflexive sensitivity of the touch in our own hands:

> When I touch my right hand with my left, my right hand, as an object, has the strange property of being able to feel too ... it is not a matter of two sensations felt together as one perceives two objects placed side by side, but of an ambiguous set-up in which both hands can alternate the roles of "touching" and being "touched."
>
> *(Merleau-Ponty, 1962/1945, p. 93)*

While there is an experiential difference between the mother reaching to touch her child and that same motherly hand being touched by her child, the hand is ambiguously capable of both gestures. It is as if there is an indistinct identity within difference as parent and child are neither completely conjuncted nor disjuncted. Instead, we have a boundary of touch alit with affectivity. The mother need not constantly look at her child, nor reflect on his wellbeing. Rather, she is in contact with him: engaged in active passivity through the passing of the day.

Both parent and child may be seen as born with this primordial sensibility. And the parent retains this infant responsivity to a world of others even after having developed an individuated sense of self. This is a human capacity that we may easily pass over, as being with others is basic to our very being as human beings (Heidegger, 1998/1946). So much so that in everyday activities, for the most part, we may spend time with others, work with others, talk with others, and so forth without pausing to have any reflective understandings about them or their understandings of us. In newborn medical care, we may witness this being together when we observe activities as gestures—breast feeding, diaper changing, kangaroo care—performed so routinely that a parent does not struggle and does not pause. But just because these activities may be performed without explicit effort and reflection does not mean that a parent is not deeply affected in a touching attachment with his or her child. Still, we may wonder, what does it mean to be with an other in ethical responsivity?

> He is a different child since the surgery. He does not have to spend all of his energy working so hard to breathe. His life is more than growing, more than waiting, more than being sick. Now, when I hold him and if he becomes unsettled, I do not just find myself trying to settle him. I find myself wondering, "What do you want? And what do you need?"

Ontologically, for the mother to be in touch in ethical responsivity is not to grasp the totality of the otherness of her child. It is not to understand his every want and unmet need. Touch instead speaks to the touch of contact: to be with an other in a feeling way (Buytendijk, 1970). And the capacity for touch allows the mother to be touched by the otherness of her child. Her way of being with him is an affective contact whereby otherness is felt rather than appropriated. To put it differently, the otherness of the child touches yet transcends the mother's touch. This touch provides the origin of the pedagogical moral response: 'What ought I to do for you?'

Although the technological may fade into the background, it is still situated in the relation of parent and child. So much so, the parental and child being may become intimately interwoven with the technomedical devices. We do not need to look to extraordinary technologies as examples. Even routine and everyday technical tools affect the relation of parent and child.

> When you get used to it, you forget about the technology. I held her this morning. It took three people to place her on me. The respiratory therapist held her tube, and her nurses took care of the wires and intravenous lines. I just held her then for a couple hours. As the monitor rang for a desaturation, and her breathing paused, I rubbed her back. The trace returned to baseline and her breathing steadied. She comes up so well. Without even looking, I hear her come up. I can feel against my chest when she needs a suction. It is a wet, vibratory feeling, then the machine rings "tube obstruction" until we suction her out. I don't really look at the monitor more than I look at her face, her body, or any of the other medical instruments. But I constantly know how the monitor reads. I am always listening for the monitor. I am always listening for her.

The opaque presence of the technology dissipates to reveal a technical touching attachment of parent and child. The Merleau-Pontyan "flesh" is the generative of what makes possible the intertwining of the sensate and the sensible: "an anonymity innate to myself" (Merleau-Ponty, 1968/1964, p. 133). In the newborn intensive care unit, the technological may become the "flesh," experientially offering parent and child an (in)tangible connection. The parent's experiencing of her child is not only biological but also technical in the mechanical reverberations of the respiratory circuit and the synthetic sound-images of the monitor screen. We may wonder what the parent responds to. The biological? The technical? Or a blending of both? Surely the way that these technologies may subtly shape the experience of self and other has ethical consequences as sensual sensibility itself is touched.

From being in touch with the child, the parent is responsive. Yet, while their bodies are together in contact, we may wonder if another meaning pervades this primordial affectivity?

Consider the mother nursing her child. In response to the mother's initiating touch, the child mouths the air searching for the breast. And when the areola is found, and a latching is achieved, child and mother are literally connected in physical contact. Still, even if this phenomenon is representative of an instinctive 'rooting' reflexive touch, it is not just the mouth closing on the breast that founds attachment. Jean-Luc Nancy points at what is transcendent to touch: "mouth slightly open, detaching itself from the breast, in a first smile, a first funny face, the future of which is thinking" (Derrida, 2005/2000, p. 21).

Perhaps ethical responsivity needs a mouth that opens to smile, or, less roman-tically, a mouth that opens to cry, to disturb the natural nesting of nursing. For Heidegger (1962/1927), it is only in special situations that the primordial mode of being with others may become interrupted, leading to instances of pause, reflec-tion, and deliberation. And during these disturbances, a parent may experience a look that looks like eye contact.

> Yesterday was the first day that she opened her eyes, and I actually got to see it for myself. Before then, her eyes had been still fused from prematurity. It happened when we were bugging her. We were taking off all of the mon-itoring stickers and changing her diaper. We were going through the motions like so many times before. Even though it has only been a few weeks, my hands have gotten so used to the routine. I found my fingers moving to slide the diaper beneath just as her nurse lifted up her legs from the other side of the isolette. And then it happened; she was looking at me. Her look became a statement, "Here I am." And at the same time, I felt it as a question, "Does this need to be done now?" I was taken aback. To see her, and to see her respond to my touch, I felt it well up inside of me.

These are occasions when a parent, in being touched by the look of her child, experiences a contacting recognition. The parent is drawn nearer: to see the child and to know the child as inappropriably 'mine.' This is the pedagogical ethical sphere between self and other. For the family in the neonatal intensive care unit, we need to wonder how a parent finds such a space when just being with the child can be technically so difficult. The survival of prematurely born infants is condi-tional on coping with the situation of infants born prematurely detached from their mother's body and needing reattachment and connection through technology and professional expertise (Landzelius 2003). Attachment between parents and their child is facilitated and emphasized while recognizing that the high technological setting of the NICU differs considerably from the usual ways an infant enters the world by being taken care of by his or her parents outside of the NICU.

IMAGING THE CHILD

The children of the NICU form a diverse population. Looking at each infant with a medical eye, the healthcare professional may get a sense of the reasons for admission. Some children are quite small, being born too early; others appear ill, from incidents incurred before, during, or after birth. Some infants appear obviously malformed, marked by physical stigmata of syndromes; others appear well but have been admitted for observation or for convalescence at the end of their hospital stay. At times, though, external signs are insufficient. A technical picture or image may be required to better perceive and understand a particular part or condition of a child.

Brain images can be produced by ultrasound, computed tomography (CT), or magnetic resonance imaging (MRI). All of these technologies have their own technical advantages and disadvantages, providing various degrees of and differences in resolution. In common, they produce images to be read by the technological eye of the expert. But what does a parent see who is shown an MRI of his or her child demonstrating brain damage?

I am not an expert, a doctor, or someone with medical training. But, I can see what is shown to me and what it means. I could see what they were showing me in the MRI of his brain. One side did not look like the other. The details of the one side were gone, and it was obvious to me that the damage was severe. It was like half the brain was obliterated into a mess of blurry white and grey. They said it showed a combination of blood and damaged brain tissue. It was horrible. Devastating. How can a child mature and develop when such a huge portion of the brain is damaged? What kind of future can he have? Now my son no longer looks the same. My son looks different after I saw that picture. It is as if the whole future, the planning, the expectations, everything has changed. He has now become that picture.

Looking at the MRI image, the father is afforded a uniquely technical view of his son. It is not an outside view in the sense of a different look at the head (and face) of his child from yet another perspective. It is not an apperception proper. Instead, there is hermeneutic quality to this view, showing an image that while being of the child is not the child as his child. While the father may be unable to name cerebrum, thalamus, cerebellum, or other structural elements of the brain, his technologized eye does see parts of the whole of his child: parts that are damaged to the point of obliteration. And he sees more than the technical image. In these damaged parts, he sees a future or more accurately perhaps he sees a broken future. His child no longer looks the same.

While the medical specialist sees with a (dia)gnostic eye, the father sees with a pathic eye. The pathic eye sees the body meaningfully: innocent and vulnerable, disfigured, and hurt. When a father or mother simply sees their child at the bedside, they do not see a body transparent and opened by the medical imaging equipment. Rather they see a closed body. They see their child in recognition of his or her familiar features. They may also see the marks, scars, and other bodily traces as memories of pain and hopeful anticipations. The pathic eye sees caringly, lovingly, or worryingly (van Manen, 1999). In contrast, the technological eye is a gnostic eye, visualizing and seeing the child's brain in constructive radiographic resolution. It is the eye that sees the world in objectifying terms of diagnostic images and pictures, to be analyzed, examined, probed, and questioned.

In seeing the brain image, the father's ordinary pathic vision of his child is disrupted. The MRI favors the diagnostic and prognostic eye at the expense of the pathic. The technical intervenes and insinuates itself as if it is filling an existential void. The eye is drawn to the abnormal, the parts or structures of injury and damage. In a sense, the child's childness has become invisible. The parent may ask: What part has been injured? What function was it responsible for? And what will be the consequence?

The MRI may lead us to ask what is wrong with this image rather than what is right, as the observer of the MRI is drawn into the medical considerations of diagnosis and prognosis. Surely if this image was just shown as a scientific image of an anonymous source then it would only be a curious technological image. Especially if the MRI only shows a picture of parts, then this is just an image of 'some' brain. But the father is shown his son's brain. An MRI picture of the inside of his son's head. And now, everything has changed. The father is not seeing an abstract image of some brain; he sees his son. This image shows what the father could not have imagined. The child's future has been put into question such that we may ask: What child is this? What child has been given as this image?

The brain image becomes the picture of anticipation. The full meaning of the picture may remain in question—the exact future of the child remains imprecise and tentative—but there is no doubt that the father's child is changed by the ethical technics of the MRI. The child is now marked by an aberration: imaged as damaged and imperfect. The image has revealed a truth of the child: His brain is hurt. In this way, the MRI technology has brought an aspect of the invisible visible child into presence.

It should be clear that the way the MRI reveals the child to the parent is different than the nontechnical way a parent comes to know his or her child. The technics of the MRI does not allow the arising of something from out of itself, as the child would naturally be given to parental experience over the course of time (*physis*). Rather, the MRI brings the child forth in the Heideggerian sense of a technical *poiesis* (Heidegger, 1977). It is a bringing-forth not of the child in his or her ownmost being but rather an interrupting and challenging bringing-forth. It is a technological *poiesis*, whereby the MRI challenges the natural *physis* of the child. In this sense, the child is perhaps revealed by the technics of the MRI to be the child of lost dreams, lost hopes, and lost anticipations (Stiegler, 1998/1994). How does the parent now gaze upon the child? And if the child can gaze back, how may their gazes meet and be sensed?

Without imaging technologies, a parent sees his or her child as he or she develops over time. The child's strengths and challenges are revealed and discovered gradually. It is only in being with a child, in spending time with him or her that a parent may come to know the possibilities of his or her child's injuries. This is perhaps even more poignant while a child is in the NICU, where the child is still less able to give of himself or herself to the world.

Consider the infant born early, not quite prepared to interact with others in the same manner as the full-term child. A premature child may provide only elusive cues and subtle responses to the caregiver. The preterm child may lack motor maturity, fluidity of movement, be unable to suck and swallow. While the term child may calm to recognizable voices, the preterm child may manifest a stilled calmness of immaturity in sleep-wake and attention-interaction states, seemingly never quite able to become fully awake and open or attentive to those around. Also, the caregiver's response may be inappropriate or unaware what the immature baby may be sensitive to. The premature child may only be soothed by the steady pressure of well-placed hands mimicking the time of being in the uterus rather than responding to the well-meaning sensitive stroking of the child's brow or hand. This relational difference may also be true of the sick or damaged child, whose interaction is restricted due to illness, brain injury, or even as a consequence of the medical technologies aimed to sustain the child in the state of sedation, paralysis, and so forth. In this context, imaging technologies afford a supplemental understanding to what is lacking in the relation of parent and child.

It is crucial, however, to acknowledge that the MRI supplement is more than an addition of information. The supplement is the image and representation of nature that is neither in nor out of nature (Derrida, 1976/1967). And once seen, it is necessarily seen. The ethical significance of the MRI supplement therefore carries the meaning of the Latin *supplementum* as "something added to supply for a deficiency" (OED). This is radical in the sense that our view of the child without the MRI image is lacking and thus deficient.

Without the MRI image, our understanding of the child's being is not whole. In other words, prior to the MRI image, the child has been given to the parent as marked by a non-presence, an undetermined condition. This undetermined condition is the unknown of expectation, opportunity, and future. Will the child excel in football or mathematics? Will she marry one day? Will he take after his father or his mother? Will she get a good education? The MRI supplies this deficiency. It transforms the expected of the child in the sense of it showing not just how the child is at this moment in time but also giving the child in advance of the child: a sense of anticipation of the future to come. In this sense, the MRI gives the non-expected of the child to fill an un-expected expectation.

And yet the role of the MRI as a supplement is ambiguous:

> It is the strange essence of the supplement not to have essentiality: it may always not have taken place. Moreover, literally, it has never taken place: it is never present, here and now. If it were, it would not be what it is, a supplement. Less than nothing and yet, to judge by its effects, much more than nothing. The supplement is neither a presence nor an absence. No ontology can think its operation.
>
> (Derrida, 1976/1967, p. 314)

There is an absolute affectivity and alterity to the MRI image, as it affects and alters the father's relational experience to his child. In seeing the image of the child, the representation, who or what is it that the father sees? For in the moment of seeing the image, the father may not see the child he had known in the initial or imagined contact. Indeed, everything has changed, as the imaged child is now the anticipated imaged child. It is both the child present before him but also the child he will come to know: to touch and be touched by. Perhaps in this way, the ethics of the MRI are revealed as profoundly affecting the parent's originary perception of the child as an other who once seen cannot be unseen.

ATTACHMENT AND RESPONSIBILITY

Sometimes I need to go home. I need a break from the hospital and everything there. But it is also hard being at home. When I am home alone, I feel
uncomfortable. The house becomes too quiet. The lights are too bright. And
there is too much space around me. It's kind of like walking into an empty
space. No matter how much noise you make, the house stays empty. Meanwhile, my ears are straining to hear him. As I think of my son back in the
hospital, I start to leak. I smell of milk. Then, I know where I should be.

Distance may afford an understanding of the potential closeness of parent and child.
The parent is with-out her child, with the leakage of milk providing an existential,
ethical reminder of the absent presence of her son. And the home loses its phenomenal meaning of familiarity. After all, how can the parent feel at home when
her child must dwell in another place? As we consider this intimacy of attachment
and responsibility, we may reflect on the phrase: "The soul is the other in me"
(Levinas, 1981/1974, p. 191).

For Levinas, the "soul" is not some spiritual entity. Instead, the soul is the ethical
event (the responsivity) of the "other" evoking concern and, therefore, formative
of "me." The parental recognition is a moral event, as the mother, in thinking of
her son, recognizes where she ought to be. This does not mean that the mother
needs to return to the hospital to respond to the ethical call. Rather, it means that
the parental response has already happened prior to and regardless of her inaction – she is existentially open to her newborn, with the capacity for such
wonders as attachment, love, and pain. The parent is touched by the ethical
demand of the child, prior to asking for something in return, feeling this otherness,
the child's need, from within. And the mother, as giver of milk, is the child's
bodily need. Such is the origin of 'attachment' as *estachier*, to "to attach, fix; stake
up, support" in need (OED).

What is problematic for the hospitalized family is that the child may need more than his or her parent can provide. In newborn intensive care, the child needs some degree of medical care. Often, the child's nutrition is provided via intravenous or silastic feeding tube. And the hospital staff regularly does the routine care of changing diapers and mixing feeds. This 'medical mothering' of need may render fecundity, parenthood in a relational sense, ambiguous.

> It feels like there are so few things I can do for Tysen. I come in every day and sit beside his isolette waiting for something I can do. I watch the nurses do their assessments. I listen to the discussions on rounds. And I busy myself rearranging the cards on his shelf. Every so often, I can help a nurse with his diaper changes or participate in some other care. It feels good to get in there, but even then, I still find without them, without the nurses, I can't take care of my own child. It's like although I gave birth to him, he is still not mine. And if he is not mine, who is his mother?

We need to consider how the development of attachment may be hindered when opportunities for a parent to interact with their child are mediated by other medical staff such as a nurse, physician, or other caring professional. How can we tactfully support parents who must repeatedly leave their newborn children day after day in a hospital nursery when they already feel distant? The children are not only born premature but are also prematurely separated from their parents for days, weeks, or months. There may well be latent existential complications to such divisions for the parent and for the child.

The NICU is not a static place; it is a place of growth and change, some expected and others unexpected.

> I came back to see him the night he had his brain bleed, and he looked different. He seemed to just stare vacantly forward. In many ways, he looked better than the day before. He was less puffy from excess fluid. His heart rate and blood pressure were reading better on the monitor. Certainly, he was on fewer medications. His color was even better than before, although they said it was from the blood transfusion. But he was different. He did not respond like my Jacob. He just lay there when I touched him. Something was very wrong with him. "Can he hear me?" I asked, "Does he know who I am – that I am here?" I really did not hear the nurse's attempt to appease me. Jacob did not look like my child.

For the parent, identity surpasses appearance, as Jacob looks better but does not look like 'my child' of expectation. He has become unfamiliar, strange, and alien. And in this moment, it is as if the strands of attachment are stretched to breaking. The child no longer reacts responsively. The touch is not the same. Perhaps for some parents, in an environment of technology and illness, it is hard to develop a sense of closeness. And so, we may worry about the ethics of situations and conditions that leave the child untouched and untouching.

> We were waiting for the results to come in, to find out whether she had Down syndrome. I did not know much about it, except that it meant she would have some problems. And yet to me, she just looked ever so perfect. She was beautiful. The doctor came in with the social worker for rounds and asked if today we could sit down to talk, probably sometime around noon. They wanted to know if my boyfriend could come too. He was working, but I figured we could put him on speakerphone. And then the time came. We sat down together with him and the social worker. And then, well, the doctor just said it, "The test results came back. She has Down syndrome." There was no apology or anything like that. Rather, he talked plainly about it: as something she had, that she will always have, something that is just part of who she is but also not all of who she is. I just wanted to hold her in that instant. To me, she looks just like her brother when he was born: pudgy cheeks and a wisp of brown hair. As I picked her up and held her close, I started crying. And then the doctor said, "She is beautiful."

In the days after a child is born, we may learn how a parent comes to know their child. For many parents, it is not that their child is born a stranger but rather there is a certain immediate familiarity, and also a uniqueness that blooms inceptually. We can appreciate how each day it can actually become harder for a parent to remember how their child initially appeared because every moment the familiarity grows stronger. And yet, in the context of delivering medical care, both child and parent are ethically vulnerable to us. We always risk disrupting a growing attachment. The parental experiences of ethical responsivity, and the formation of attachment, are saturated with complex meanings. Hopefully, as we gain insight into this phenomenon, we can better understand the requirement of nurturing moments of togetherness between parent and child, especially when medical care is happening.

The profundity of parental ethical responsivity speaks to the significant position parents play in making decisions for their children.

> In the passage from birth to death, the newborn baby has been incorporated into the social world through the mediation of neonatal medicine, but these intense and primary ties have to be estranged and severed at the time of their departure.
>
> *(Seo, 2016, p. 568)*

We understand that the experience of ethical responsivity and formation of attachment shifts, changes, and develops in the formation of the parent-child relation. So, we may wonder about the significance of this metamorphosis. Is the decision a parent makes for an as yet unborn child ethically different from a decision made for a newly born child? How is it that some situations may ease or complicate a decision? These questions deserve consideration. As the ethics of birth unfold in a newborn medical nursery, we may have to consider the existential consequences of an increasingly technicized life. After all, it is from this techno-medical place that the symbiosis between a parent and child must take root and grow.

SECTION III
Ethics and Decisions

SITUATING DECISIONS

I woke in the middle of the night with a tightness of my belly extending to soreness of my back. Shifting from side to side in bed was not helping, so I got up to get a drink of water. But before even making it to the kitchen, I felt the wetness between my legs. That is when the cramping began. As I called for my husband, there was nothing but confusion. We hastily grabbed clothing and searched for the car keys. All I could think between the cramps was that this is too early. We drove to the hospital in a daze. I was whisked onto a stretcher in the emergency room. "There is still time," I heard a nurse say, "This doesn't need to happen here." I was rushed to labor and delivery. There were so many people coming in and out, calling things out in that room: "How far along is she? Did anyone check how dilated she is? Is NICU on their way?" I know different people introduced themselves to me, but, honestly, all I was thinking was "it's too early!" I am only 22 weeks along. One doctor was trying to explain to my husband that the baby is probably too small to survive even if we try. I could tell my husband was trying to understand that the doctor was looking for agreement. All I could feel was cramp after cramp. I could not hold my legs together. I had to push. And then it happened. The baby came out. They took her, probably for only a few minutes or so, and then brought her back to me. "She is just too small. There is nothing we can do."

We need to acknowledge the realities of decision making in neonatal-perinatal medicine: Most parents do not plan to have a baby born needing intensive care. Conversations often unfold in times of commotion, urgency, and duress. Families may know little about what decisions are available to them, or lack the knowledge that professionals feel is needed to make a truly informed choice. And families and healthcare team members may have strong yet varied convictions about appropriate courses of moral action.

The circumstances of these ethical realities of decision making do not only precede possible decisions; they constitute new decisions as the decisions unfold. The problem is, however, that many bio-ethical perspectives distance themselves from the experiences of decision making, as the perspectives treat decisions as dilemmas to be analyzed or problems to be solved (Ladd & Mercurio, 2003; Leuthner, 2001). Different conceptual models have been developed to provide structured frameworks for determining and evaluating moral issues, mediating between abstract moral theory and healthcare practice in its universality (Beauchamp & Childress, 2001).

In the neonatal-perinatal medicine literature, there exists an abundance of argument-based publications aimed at elucidating, interpreting or criticizing ethical concepts to justify actions (Cavalo et al., 2020). One such concept is personhood, defined as the capacity or potentiality for self-consciousness, rationality, and choice (Finnis, 1994; Tooley, 1972). But the project of bio-ethics turns abstract when debating at what point personhood comes into existence, before or after birth. With regards to such an issue as personhood, Jennings (1990) writes,

> These are not basic terms with an identifiable meaning … these concepts are woven into and take their meaning *in situ* from the medical nature of the newborn's condition, his or her response to treatment, and the perceived personality traits and characteristics that caregivers (especially nurses) read into the behavioral responses of the newborn over time.
>
> *(p. 270)*

Theoretical and conceptual language tends to lack ethical texture and therefore risks reducing the complexity of experience into what can be easily conceptualized, in turn obscuring and obfuscating the need to understand the lived experience of decision making. When we are confronted at the bedside by actual situations needing decisions, we realize just how lacking and limited the argument-based literature is. We need a discourse that allows the prose to be 'sacred' and 'dignified,' that gives voice to our intuitions that some things have inherent value, even if they are resistant to conceptualization (Dworkin, 1994).

For the above mother, only 22 weeks along in her pregnancy, we need to acknowledge that her decision emerges from a body in labor. Some might say that she had the choice whether to come into the hospital in labor while still so early on in pregnancy, that she could have remained at home and let nature take its course. But we ought to concede that some situations are not seen as choices in moments of alarm, commotion, and panic. We also should appreciate that although the healthcare team is probably doing its best, it is hard to have a conversation through the pains and efforts of labor. Moreover, even if delivery is not forthcoming, it may be challenging for parents to enter into decision making when the desire is quite simply for the pregnancy to continue, for the baby to survive, or for everything to somehow end up ok. As the cramping intensifies, we should not be surprised that options, alternatives, and choices are possibly experienced as lacking. And when the baby is born, we may wonder whether it is the healthcare professionals who ultimately need to take the weight of the decision in such a situation: "She is just too small. There is nothing we can do."

There are other situations,

> The doctors and nurses were trying so hard: trying to get IV lines in, manipulating his breathing tube, adjusting medications. There was not a moment when someone did not have a hand on him. I am pretty sure the medicines and machines were maxed out. We so wanted him to live. They were doing everything they could. But it did not work. In the end, he did not die alone. It is just that we never really had a chance to hold him.

There is considerable literature on the roles of families and health professionals as decision makers, recognizing that over the past 50 years the culture of neonatal-perinatal medicine has shifted from one of parent-child separation to family-centered care and from medical paternalism to shared decision making (Barry & Edgman-Levitan, 2012; Davis et al., 2003; Harrison, 1993; Philip, 2005). Parents are charged with relative authority as prime or surrogate decision makers to determine what is in their child's 'best interest' (Buchanan & Brock, 1989; Kopelman, 2006; 2007). Their autonomy ought to be respected, as we recognize that best interests are not always clear, particularly when we are confronted by moral boundaries of sanctity of life, quality of life, and professional responsibilities (Gilmore, 1984; President's Commission, 1983).

> When parents make decisions about the treatment of babies who are very badly damaged, then, they do not and should not decide on the basis of some impersonal and impartial best-interests standard. They do it out of the relationship that holds this particular baby inside the value structure of these particular parents.
>
> *(Verkerk & Lindemann, 2014, p. 504)*

However, even with such shifts in culture, we realize that closeness may be compromised by medical efforts aimed at sustaining life. There are ethical moral costs to intensive care that we need to apprehend.

Where theoretical accounts are lacking, a phenomenological perspective gives priority to understanding the lived experiences of decisions. The phenomenal meaning of a decision lies neither wholly in the subjective nor the objective realm but rather in the manner in which decision makers are confronted by their situatedness. For philosophers such as Emmanuel Levinas (1969/1961), Knud Løgstrup (1997/1956), and Bernard Waldenfels (2011/2006), the ethics of a decision are situated not in abstract theories but in the lived experience of the face-to-face encounter. Ethical moral situations emerge out of the immediate, pre-reflective involvement we have with those others around us (Ricoeur, 1966/1950; Sokolowski, 2017).

In comparison, a casuistry approach to ethics involves considering morally perplexing cases by placing them in juxtaposition to "well-described real or fictional situations" (Jonsen, 1980, p. 159). The problem with such an approach is usually similar cases are identified based on what medical features are shared. In the practice of casuistry, a moral understanding does not precede but follows the study of particular cases, such that the theory yielded is a function of the cases selected. Ethical moral issues themselves may be unvoiced, remaining silent yet present. Still casuistry is more than talking about cases: it must have the "quality of moral discourse" (Jonsen, 1986, p. 71). As such, even those reasoning through ethical moral dilemmas, appealing to paradigm cases, need to consider the unfolding of the experiential realities of those involved in the decision. In neonatal-perinatal medicine, attention needs to be directed to the child, both how we experience the child in his or her childness as well as the child's possible life experience in itself.

> As we were talking, the doctor said that they would support us in our decision, that it was a good decision, that many parents and doctors would also make the same decision in our situation. That was a great relief, because some of our friends and family, well, they said that we should not give up hope, that we just had to give it more time, that the doctors surely could do something if they tried. But those people, well, they were not here. And this was not their child.

Decision making for someone else cannot help but be complicated in the case of a variety of opinions in the context of a plurality of societal values, just as decision making may also be fraught with uncertainty and unpredictability (Lantos, 2001; Mesman, 2008). As families are recognized and supported in decision making, for some, patient-centered care has become synonymous with family-centered care (Committee on Hospital Care, 2003; Haward et al., 2017). Conflating these notions, however, is fundamentally problematic. Family-centered care points to treating a child within the context of his or her family; patient-centered care means placing the child's interests upfront. Divergence, disagreement, or impasses are always possible in matters of decision making.

For neonatal-perinatal medicine, decisions unfold in varied and contingent contexts—from before a baby is born to the cradle of the NICU. The decision maker is generally the parent confronted with the responsibility for their child. And yet, parents are not alone in their decision making. The presence of multiple medical caregivers (nurses, physicians, and so forth) and the technomedical environment itself may complicate the nature and quality of contact between parents and their children (van Manen, 2012a; 2012c).

And so, there is value to reflect on the meaning of deciding for a child: Are all moral decisions truly decisions in themselves? How is the child given in decision making, recognizing that he or she is coming into being as a child? How does the technomedical relationality of neonatal-perinatal medicine affect the decision making of families? What is the lived meaning of decisions as they unfold? These themes are pursued in the next chapters.

DECISIONS WITHOUT CHOICES

In the NICU, the newborn child is generally considered the central concern to the ethics of moral decision making. The vulnerable child, who is yet without autonomy, relies on his or her parents and the health professionals for decisions regarding medical interventions (Miller, 2006). Sometimes it is the outcome of an investigation that announces the necessity and possibility of a decision, such as a head ultrasound showing severe brain injury in an ill infant. Other times, it is the constellation of intensive therapies utilized in the deteriorating condition of a child that signals a decisional juncture. In such moments, healthcare professionals may call on parents to consider options—the possibility of a decision. The alternatives may include initiation, limitation, or withdrawal of medical therapies. For parents, the choice is not always seen.

> The doctors kept on asking us about withdrawing care. We felt pressured to decide, almost hounded, to take Sam off life support. It was as if they thought that we did not get it. But we knew that he could be severely disabled, that his chances were so poor. Yet, how could we kill him? How could we have a part in ending his life? We avoided the staff to avoid the discussion. We avoided coming in to see our son, just to avoid being confronted with the predicament of having to face some kind of impossible decision. We just wanted to let him have a chance. If he was to die, he would die on his own. We did not want to take his death away from him.

From the healthcare professionals' catechizing call "to take Sam off life-support" and the parents' inability to respond in consideration—"how could we?"—we may discern that sometimes a moral decision is experienced as a choice without choice. Are such choices in themselves decisions? And if not, do such differentiations or distinctions matter?

We regularly use terms such as want, wish, or desire in our articulation of decisions. Nonetheless, when we are content with a family's decision, we do not necessarily explore its motivation nor necessarily differentiate what is chosen from what parents ultimately want from their decision. Perhaps it is in part because we intuitively recognize that many of our actions occur without reflection. For example, without deliberation, a mother may find herself reaching to touch her daughter after she is born; without forethought, a father may find himself talking to his son at his bedside; or, without premeditation, a parent may find him or herself looking at their child in his or her childness. And of course, some meanings are expressed in discrete actions. For example, a parent may find him or herself hoping for their child to thrive and leave the NICU, desiring that their baby will be able to feed orally, or quite simply wanting more time with him or her. None of these wants are necessarily directly fulfilled by explicit decision making.

We need to acknowledge that the notion of choice seems to imply some degree of voluntary decision. But there is a contradiction to the power of choice: It is not fully free. To decide for an other means a responsibility for the 'other' and for the sake of the other. This other (the child) is experienced as making a demand on 'me' (Levinas, 1969/1961). We could say that there is accountability to a moral decision that emerges from the demand and claim of responsibility. The decision presents weight, burden, and charge. As the parents are being asked to decide for their child, "to take his death away from him," we may wonder how this choice is within their control. Is deciding to take part in ending their child's life a choice? Are the health professionals asking too much?

> It was just too much, too much for a child. We would be putting him through too much pain and, really, not giving him a life. There would be no guarantee that he could feed like other children, that in the end it would all be worth it—what he would have to live through. Taking him home. Introducing him to our family. These are things we can do for him, even if that means he will only be with us for a short time. This is the life we can give him. Surgery just is not an option.

The bio-ethical literature often views decisions of initiating, limiting, or withdrawing of medical interventions as conceptually motivated based on foundational perspectives of 'life value.' Distinctions are made between 'vitalism,' 'sanctity of life,' and 'instrumental value.'

Vitalism holds that life in itself has an ultimate value no matter what form it takes, such that medical interventions should be pursued to preserve life, regardless of the unfavorable consequences of the interventions. Certain religions, cultures, and perspectives are implicitly or explicitly vitalistic.

Alternatively, a 'sanctity of life' doctrine, while recognizing the intrinsic worth of life no matter what form it takes, tends to make distinctions between actions which preserve compared to actions which end life. Those endorsing either a vitalistic or sanctity of life perspective would be against intentionally ending life (i.e. active euthanasia); but, limiting the use of medical interventions that have the possibility of prolonging life (i.e. passive euthanasia) would only be potentially supported by a sanctity of life perspective and not a vitalistic view. In other words, the critical theoretical bio-ethical difference between these perspectives inheres in the 'inviolability of life' rather than a 'valuing of life'; and as a result, expresses different imperatives for action.

An 'instrumental view' sees life as of value in so far as it allows for 'goodness' (i.e. valuable experiences, meaningful relationships, or other good). From an instrumental view, life is foreseeably violable, such that it would be moral to support discontinuing medical interventions if a child is no longer supported for a life with quality from it. From this perspective, end-of-life decisions are shaped by at least three considerations:

(1) The medical intervention does not improve the situation; it isn't effective (this is a question about medical futility); (2) it might be effective, but it would cost the child and perhaps others too much to get to the improvement (this is a question about disproportionality); (3) the ultimate quality of life achieved would be very poor (Verkerk & Lindemann, 2014, pp. 501, 502).

For the above anecdotes, we can recognize how the moral choices of the parents, what they perceive as ultimately good or bad for their child, can be conceptually analyzed along with determinations of vitalism, sanctity of life, and instrumental value perspectives. However, we need to acknowledge that such abstract, theoretical classifications pass over the experiential lived-through sense of ethics arising in a situation lacking choice. In other words, conceptual theory has the tendency to abstract or otherwise pass over the ethics of how we may experience being constrained, limited, or otherwise affected to be without a choice in deciding for another person.

> Dany is dying. The doctors keep on saying they are looking for other options of treatment, that they have not given up. But we know her life is coming to an end. Still we will not stop. Each day we come into the NICU and sit at her bedside: we hold her and talk to her, we talk about how she might be feeling, we go through the motions of pumping breast milk and changing diapers, we have conversations with her doctors. We know that she does not have much time. We have our time with her, and she has her time with us. There are no choices to make. We are going to be with her until the end.

There is no deliberation or debate for a decision without choice. It is as if the decision never arose because no viable choices exist as possibilities. Such an understanding of choice does not interrupt the temporality of life, even when this life is coming to an end. It is not actually a question of vitalism, the sanctity of life, and instrumental value: "There are no choices to make."

> The doctors all assembled in a room with us to talk about the situation. We could tell they were uncomfortable. None of them seemed to want to be

> there. After moments of silence, it was the surgeon who spoke up. He told us
> that surgery was possible but she might not live through it. There were con-
> siderable risks, and it was very likely that if she survived the surgery she would
> require additional operations. It was likely that our daughter would never
> leave the hospital even with this intervention. "But," he said, "It is your
> decision." I did not see an option. I am her mom. And she needs the surgery.

When health professionals present medical interventions as options of choice,
treatments are marked by traces of possible benefit. In other words, a treatment
foreseeably has a potential advantage for the patient—otherwise, there would be
no option to present. Should it be surprising that parents choose surgeries or other
interventions with considerable risks or of considerable invasiveness? Or that parents may find it difficult to say, "No," "stop," or "discontinue" a therapy? What
choice are we offering to them?

There is also a sense of choice as far as what is experienced as what 'I' ought to
do as 'her mom.' Virtue ethics does not go deep enough to show us the ethics of a
decision of 'what a good parent would do.' The point is we need to recognize in
neonatal-perinatal medicine that parents are confronted with making decisions, not
in abstract or hypothetical roles but rather in situational realities of being a parent
for their particular children.

> The doctors wanted to call a meeting to discuss taking Johnny off of
> the ventilator. The injury on the MRI of his brain was so severe. They told
> us that. They told us he could not survive. But how could we stop? He is
> our son.

Sometimes there is no choice option that aligns with a parent's values. While it
is possible the choice may be constrained by religious or cultural convictions,
expressing senses of vitalistism or sanctity of life perspectives, it may also happen
that an offered option is not what a parent can envision doing to his or her child.
The decision to withdraw treatment or not pursue an intervention may be
incommensurate with what a parent may feel that he or she can do 'as a parent.'
There is no choice. So, it may happen that in such a moment, when we think we
are dealing with a moral decision, for the parent there exists no real choice except
to continue on: to let the child have a chance at life, however that life might
unfold. For other parents, the decision may differ such that the only choice is a life
home from the hospital, accepting that a shortened life is the only meaningful life
that they can offer their child.

From a rational perspective, this kind of 'non-decision' can be seen as a decision
as well. But the point is that health professionals need to understand that this kind
of non-decision decision is a possible situation that parents may encounter. By no
means does this acknowledgment resolve the potential moral impasse or distress
that may occur when parents and professionals truly disagree on what is 'good' or
'best' for a particular child. Nor does it resolve the health professional's need for

consent from parents to support a decision that they wish to pursue. Instead, understanding decisions as non-decisions may prompt professionals to reconsider their approach to particular ethical moral situations. After all, in situations where the clinical-parent relationship is fractured, the "relationship is needed more than ever" (Marron et al., 2018, p. 169).

For non-decisions, there may be value in shifting the conversation away from 'what ought to be done' to explore the meaning of the current course of action: letting a child 'have a chance' and certainly the hopes, dreams, or other wants that would be good for the child to have a chance at. Similarly, there is value in exploring the meaning of 'giving a child a life' as far as what constitutes a life worth living, or the sense of 'being well' as concerns what wellness we are trying to achieve for a child. And indeed, health professionals need to consider how their words and actions may lead parents to experience the reality of such decisions so they can appreciate how medical decisions are affecting their children. We need to support parents in such a way that they can be present for their child in his or her situatedness.

LOOKING FOR WAYS OUT OF DECISIONS

There are ethical situations that seem to demand information. The decision is not resolvable in the moment but instead requires reasoning, calculation, deliberation, or even meditation. We can appreciate how many healthcare professionals are comfortable with such decisions as they recognize that parents want information to inform their decision making. The parents are looking for evidence to clear away uncertainty.

> The choices were that we could keep her on the ventilator without giving her dexamethasone, knowing she was going to deteriorate slowly; that we could give her the dexamethasone and see where that takes us; or, that we could withdraw treatment. The first was not a decision for us. We did not want her to suffer. We so wanted to help her, but at what cost? I remember taking a walk with my husband after talking with the doctors. We talked about quality of life. We knew that while the dexamethasone could be good for her lungs, it could be bad for her brain. We talked about what we were prepared to live with. We knew that if she was severely disabled that we would be unable to handle that. We did not want that for our child. We talked about what we knew and what we didn't know—the uncertainty of it all. It really felt like there was no clear right answer. And we felt like we did not have time, there was no time to make a clear decision. Time was pressing; we almost decided to withdraw treatment. But we really did not want to let her go. Perhaps we were kind of looking for an excuse to give her the dexamethasone. As we were walking back, still unsure, my husband brought up "well, what about if we ask for a head ultrasound to confirm that she does not have a brain bleed?" 'Cause if she had a brain bleed, we were not going to go on with it, 'cause then we would know that there was brain injury. So, we said if there is no

brain bleed, we will go ahead with it. We felt that this would be a way out of having to decide.

As the parents try to work a way out, the sense of the need for a moral decision may be forestalled. We realize that the 'ends' of decision making are not necessarily chosen—"We did not want her to suffer"—but rather express the contextual 'view' in which the decision may be made.

The choice, to give or not give dexamethasone, *emerges* to 'mediate' the ends of the decision (Sokolowski, 2017). Put differently, the existence of a potentially therapeutic medical intervention creates the possibility for a choice to be considered should the parents be able to envision employing it. If a choice is realized, the want (for her to not suffer) is modified to categorical form (to suffer or not to suffer). And yet, we should recognize that such categorical distinctions may become distanced, disentangled, or even distorted from the original decision for her not to suffer as other meanings of choice are explored, such as disability (to be or not be disabled), quality of life (to have or not have quality of life), and so forth, which may or may not correspond to a parent's sense of suffering.

In many ways, we may think the reasoning of decision making is congruent with a consequentialist perspective: weighing the risks and benefits of treatment such that the consequences of an action are the ultimate basis for judging the moral appropriateness of the action.

> In these theories a particular act is morally neutral until its outcome can be evaluated; the act is thus seen as something chosen in view of its outcome, and the outcome makes it a truly moral act ... The act itself is not good or bad; something beyond the act, something other to the act, makes it so.
>
> *(Sokolowski, 2017, p. 54)*

However, the ethics of a decision do not narrow to the consequences of actions (consequentialism), just as they do not rest purely on what obliges an act (deontology), the character of the decision maker (virtue ethics), comparisons to other cases (casuistry), and so forth. After all, what the parents look and long for is not just the raw information of how a drug, such as dexamethasone, may affect the lungs and brain. Nor can we even reduce the parent's looking and longing to their desire for their child not to suffer. Instead, the parents look to anticipate, avoid, or resolve the inevitability of an unavoidable course of action or inaction affecting their child from their decision.

> The doctors left us to the choice. We could remove the breathing tube now, and his time with us would end. He would die. Alternatively, we could wait. His brain might recover enough for him to breathe on his own, recognizing that the severe injury to his brain would not heal. He would be severely handicapped. We had to make a decision we could live with. But we just

could not bear it … to be the ones responsible for his death. And so we asked, "What is a reasonable time to wait?"

There are differing moral perspectives regarding what makes a 'good' compared to a 'bad' decision. From a phenomenological perspective, the morality of a decision ultimately is reflected not merely in the consideration, deliberation, or wondering about what is the 'right' decision; but instead in the understanding of choice in its 'rightness' as it unfolds. Understanding is not theoretical in the sense that it can be conceptually reduced to what can be explicated and explained, whether it follows a series of principles, consequences, or abstract notions. If this were the case, then most people could probably be convinced about the 'rightness' or 'wrongness' of decisions by appealing to pure reason such that situations of decisional conflict would not occur. Understanding instead speaks to how the decision experientially 'appears to me' as being 'good' or 'bad' to those 'others' whom the decision affects.

For parents, the existential meanings of choice options are crucial—to cause our child suffering, to let our child go, to be responsible for his death, or to carry on with our child. This is vital to understand, particularly in the context of the NICU, where death most commonly occurs following a decision to limit or withdraw medical interventions (Dupont-Thibodeau et al., 2014; Lago et al., 2008; Ryan et al., 1993; Sands et al., 2009; Verhagen et al., 2010). The morality of a decision expresses the 'choiceness' of the choice to the extent that the choice resounds with a parent's situated sense of the rightness of the action or inaction for them and their child. The morality of the decision therefore transcends consequentialistic and deontological perspectives that find no ethical difference between pursuing and discontinuing medical therapies; there are different textures of meaning to initiating, limiting, and withdrawing interventions (McHaffie, 2001; Pinch, 2002; Wilkinson, 2013; Wilkinson et al., 2014). The parents may wonder: What will life be like for our child? What will life be like for us? What can we live with?

As we unfold the ethics of decisions, we begin to realize that the 'interests of the child' are not necessarily viewed in isolation from 'self interests' or the 'interests of others.'

> Most parents from time to time view their children as extensions of themselves. This attitude is not always bad; it can motivate parents to promote their children's interests. But it can also blind them to the distinction between their children's interests and their own.
>
> *(Parens, 2009, p. 23)*

And yet, from communitarian ethics, we need to recognize the child exists as part of a family, including its relationships, responsibilities, and values (Callahan, 2003). Ethics do not blind so much as they constitute a different way of seeing.

As the decision expresses a responsibility towards the child, we may realize that ultimately decision making (expressing a looking for a way out as the decision moment) is not necessarily experienced as a choice. Instead, surrendering to the 'fateful' outcome of an investigation, the head ultrasound, or perhaps appealing to someone else's judgment of 'appropriate' waiting time defers the decision. Maybe the ultrasound provides the parents with a different picture, a hopeful situation where everything may still turn out all 'right,' by showing them a child who has no definitive evidence of brain injury. But the ultrasound has turned into a fateful coin that is tossed to find a way out of an impossible choice. In comparison, to wait may only forestall the decision, whereby an appeal is made to time itself to resolve the necessity and weight of the decision.

The problem we recognize in practice is that such turns to medical investigations may express an appeal to a source of information that ultimately is also imprecise, possibly vague or filled with uncertainty. Ultrasounds, while useful in showing blood in the fluid-filled spaces of the brain (i.e. the ventricles), accumulation of fluid in the ventricles over time (i.e. hydrocephalus), or large bleeds into the sub-stance of the brain (i.e. intraparenchymal hemorrhage), by no means show the *absence* of brain injury (Mirmiran et al., 2004; Roelants-van Rijn et al., 2001; Vollmer et al., 2003). And waiting to make a decision may also impart pain, burden, or distress on a child as medical interventions are continued without a clear goal in sight. Although the decision to turn to investigations or time may be rea-sonable, the rationale may be somewhat lacking.

Parents may need support in making decisions.

> They talked about the pros and cons of the different surgical approaches. It seemed that while one was more risky, if it went well, our son ultimately could have a more secured future. But I am not a doctor. I cannot make these decisions. The only thing I can say is that I want the best decision for him. I want the decision that they would recommend.

Clinicians need to discern between parents' informational needs for deliberation and their desires to be involved in making the decision, broadly defined as "pro-blem solving" and "decision making" (Deber, 1994, p. 423). Some parents may want information but not be involved in the decision; some parents may want both, or some may want neither, relinquishing the entirety of the decision to the expert medical team (Bogardus et al., 1999; Caeymaex et al., 2001). Consequently, as health professionals, we need to be able to work out the difference between wanting and not wanting to be involved in a choice, yet nonetheless having a desire as far as what is in a child's interest. However, we also need to recognize that healthcare professionals are not necessarily skilled in judging parental preferences for involvement in decision making or determining parents' desires to be achieved from the decisions themselves (Deber, 1994; Doron et al., 1998; Godolphin, 2003; Payot et al., 2007; Perlman et al., 1991).

Finally, we realize that the healthcare professional needs to consider his or her own ethics and professional integrity. Parental authority is not absolute, even if the rights of parents to make difficult and life-altering decisions for their child should not be infringed on without good reason. Should we not be uncomfortable if a decision rests on an inaccurate understanding of the validity of medical information? Should we feel assured that more time is reasonable to give the parents if it is the child who we are ultimately responsible for? Or should we simply be satisfied that a decision falls within the consideration of reasonableness, recognizing that what is the child's actual best interest may be hard to determine?

THINKING AND FEELING THROUGH DECISIONS

In deliberating a decision, rational calculation may not resolve uncertainty. Instead, one may need to "feel through" what is to come.

> They told us that it would be for the best if we were to discontinue life sup-
> port. Our answer was "No." We were just not ready. It was a shock. And we
> had not had time to discuss it. We felt so rushed, pushed to come to a deci-
> sion. They wanted an answer. We went back to our room. I was thinking, he
> is my little boy. I pictured him in the isolette as I stared at the pictures of him
> on the table. He was with us in that room. We searched the Internet, pulled
> everything up on the computer on what he had and what the outcome could
> be. We looked at the worst of the worst, given his brain injury. We asked
> ourselves, can we handle it? It is going to be a daily-life thing for us, and if
> that is how he is going to be, can we deal with it continuously day-in and
> day-out? We started talking about different home situations and stuff—even to
> the point of bedroom space and such—room colors, pictures, bedding. We
> were putting him at home in our thoughts. I don't think we were deliberating
> and weighing medical alternatives; it was just, can we do this? And it was at
> that point, as his parents, we knew we could do anything, that, no matter
> what, we could handle it. And if he wanted to go, he would have to go on his
> own.

As the parents pass from hesitation to choice, the moral decision may be experienced not so much as a deliberative affair of weighing alternative benefits and burdens but rather an asking, "Can we do this?"

Paul Ricoeur (1966/1950) wrote, "There are no decisions without motives" (p. 66). But motives are not simply causes, intentions, or drives. They exist in the sensible core of the moral decision regarding 'my' child: the child with hair that has never been brushed; the child with skin that has not seen the sun; the child who although being marked by profound illness and scarred by medical interventions is still innocent to the parent. The moral decision only has sense, only has meaning in relation to the child as my child. As such, the answer to "Can or should we do this?" may not merely be a response to a question of feasibility. Instead, the answer involves feeling the anticipated weight of the decision "day-in and day-out" for this child. It is the eidetic meaning of ethics of the decision as founded in the never-ending experience of the appeal of this child's being and this child's face.

Feeling through the consequences of a decision may result in consenting to the appeal: "putting him at home in our thoughts." This consent is not simply the outcome of a deliberated judgment. Consent is in "the act of the will which acquiesces to a necessity" (Ricoeur, 1966/1950, p. 341). As the parent feels for him- or herself through the daily traumas of the consequences, an initial decision may unravel from what was initially perceived as an option.

There are other parents and other decisions,

> They have asked us to start thinking about a G-tube because Sophie is not progressing in feeding. It seems like she is going to need tube feeds for months, if not years. I guess it is not so bad. The feeding tube will be under her clothes. It can be hidden. And if she does well with her oral feeds, the tube can be removed. It does not have to be a permanent thing. It is just, by accepting that she needs a tube, it is like we are somehow giving up on her. Going home with a feeding tube is not what we were working towards.

As something different than expected, the decision needs to be felt, mulled over, or otherwise tentatively experienced in its capacity to be lived through. This should not be confused with fantasy or speculation. Thinking and feeling through a possible decision means anticipating the decision in its eventuality, such that it becomes possibly constituted yet also transformative of our expectations. This is the space that needs to be traversed in thinking and feeling through a decision.

We need to distinguish between feeling through a decision and being emotionally involved in decision making. In response to the concern that emotional decision making is irrational, leading to poor decisions, many have championed emotions to be critical for enabling sound decisions, arguing that emotions help to refine and clarify moral values (Godolphin, 2003; Janvier et al., 2012; Janvier et al., 2014; Lerman et al., 1993; Slovic, 2001; Walter & Ross, 2014). However, to speak of emotions in this manner for decision making is misleading because emotions themselves arise from how we are engaged in a decision, who we are deciding for, and what we are deciding. While we should acknowledge that emotions in bioethics are important, it is not the emotions themselves that deserve our primary attention. Instead, we need to recognize the ethical moral experiential realities of decisions. If our discourse of decision making becomes overly theoretical, we risk reducing the decision to a situation where we have lost touch with its ethics.

This is now the sixth time she has failed extubation. I know what is coming next. They are going to talk about a tracheostomy. It is a conversation we need to have. When I arrived to the NICU today, I heard she had another night of not really sleeping. Constantly shifting in her bed, writhing against the tube, it is like she is living in a constant struggle. So, this morning she was just completely sedated, made to rest by the medication and maybe also her pure exhaustion. But I cannot help wonder if we should wait just a little longer for a tracheostomy. If we should try one more time to take the tube out. I really do not want the surgeon to carve a hole in her neck, to be the one responsible for it. With a tracheostomy, the next months and maybe even years of her life will be different. When we will finally be able to take her home, we will be bringing the intensive care home with us: suction machines, a ventilator, feeding tubes ... even nurses living in our house. This is just not the life we had planned on giving our child.

The decisions that parents make can be considered as felt through 'parenting decisions' that shape the childhood they afford their children. Families are the places where cultural traditions are taught, practiced, and internalized (Nelson, 1992); and perhaps more importantly, we hope that parents are uniquely situated and prepared to make decisions that serve their children's 'best interest.' However, in our pluralistic society, parents may differ in their sense of what constitutes 'good' parenting choices. Just consider the varied responses to such parenting questions: Is it appropriate to let a child cry him or herself to sleep? Are punishments appropriate for bad behavior? What expectations should we place on children as they mature? We are accustomed in our society to respect that parents have much autonomy to make decisions for their children, even though newborns and other young children may hold values that differ from their parents when they mature (consider how many adolescents and adults look back at their childhood criticizing their parents' decisions).

Where neonatal-perinatal medicine is unique is not only that families are given considerable decision-making authority as decision makers but rather that they are making decisions for new-borns as new-parents. Parents are making decisions for a child that they may have yet to name, that they are still getting to know, that they have not yet taken home. Even if the parent has other children, the relation to this 'new' child is developing in its own way. We need to respect the becoming of this new parent, whereby the parent comes into being a newborn parent from conception, through pregnancy, to the cradle of the NICU.

For moral problems in need of deliberation and information, the healthcare professional needs to attend to supporting the parent and child. Since this process requires not only a 'thinking through' but also a 'feeling through,' professionals may need to do more than meet informational needs of parents in a caring and compassionate way. Ethics, after all, requires not just the weighing of benefits and burdens but also the resolve from parents to ask themselves: What do we feel comfortable with? Does the decision feel right? And, ultimately, can we live with it? This manner of proceeding does not necessarily require a moral decision apart from the 'best interests' of the child. Instead, it may situate the decision relationally in the parent's encounter with his or her child as the parent asks: What ought 'I' to do for my child? Deliberation or calculation may not resolve uncertainty. The healthcare professional should therefore recognize that it is not only information that informs a decision.

And yet, the justification for parental decision making is contingent and revocable. Healthcare professionals have a responsibility to not merely accept any decision a family wants without also considering the impact it will have on a child. Professional practitioners, too, are a part of the broader societal community charged with advocating for children to ensure that decisions fall within the range of societally 'acceptable,' 'good enough,' or 'reasonable' actions (Rhodes & Holzman, 2014), or quite simply decisions avoiding harm (Diekema, 2011).

DECISIONS AND INDECISIONS

Sometimes, a situation requiring decisions may be experienced as an ethical moment that places the parent in a predicament of impossible responsibility.

> We were asked to make a decision about Isaiah. We needed to decide whether to carry on, or whether to stop. They told us that if he survived he would almost certainly be severely impaired: never walk, never talk, and likely never see. He would not share our world, at least in the way that we live it. Thinking of Isaiah, I thought of my other children. I remembered their faces when they were young. I thought of the dreams we had for them. I thought about what continuing on with Isaiah would mean for them. We would have less time for them. There would have to be sacrifices. Sacrifices that I would never want my children to have to make. Still, Isaiah is also my child. His face is also my child's face!

What ought to be decided for Isaiah? How much burden is too much for Isaiah's siblings to bear? How many and what kind of challenges are too demanding for any child?

The etymology of "decision" means "to cut off" and derives from the Latin *de-caedere* (OED). Cutting, inevitably, divides one option from another, with each division not only creating cuts between different choices but also potentially between responsivities, demanding sacrifice.

Dan Brock (1996) conceptually refers to moral considerations that fall outside of the patient's "best interest" as "non–patient-regarding grounds" (p. 608). In neonatal-perinatal medicine, we recognize that in addition to our patients it is the families who are most affected by decisions, as they bear the psychological effects, social repercussions, financial costs, and other consequences of the decisions that are made. Given the impact that medical decisions may have on an entire family, it seems only appropriate that families have a discretional role in making decisions. Brock's "non-patient-regarding justifications" are stirred in daily family life as parents judge what is best for their family. However, bio-ethicists tend to insist that a decision should fall within an acceptable range of options, primarily guided by the child's interest rather than a family's own needs. In other words, external justifications are generally felt to have less moral weight than patient-centered justifications.

The language of non-patient-regarding grounds poses a problem. It abstracts the ethical moral language of decision making. In the father's ethical response to his child's face lies a profound sense of responsibility. The ethical appeal of each child transcends rational grounds; the father is claimed by the faces of all of his children. He knows himself to be responsively responsible for Isaiah, and he is also responsible for all of his other children. The responsibility does not originate from himself but from the others: his children. The conceptual rationality of non-patient-regarding grounds misses the ethical predicament that the father finds himself in. It does not point to the father's reality that the significance of the decision may be felt in the experience of being held in indecision: The father cannot satisfy the demands of responsibility for all his children at the same time. As Gayatri Spivak writes, "ethics in an experience of the impossible" (1995, p. xxv). Perhaps all the father can responsively and responsibly do is defer the decision as indecision, to endure the indecision. It is the impossibility that keeps decisions open.

Bearing the ethics of moral decisions may wound or heal, scar or celebrate, or otherwise profoundly affect us. It is as if a decision cannot help but persist as chosen through time, as the choice transcends the unfolding of the decision.

> The I, if it experiences the call, has to recognize the absoluteness of its precedence; and practically, the I cannot decide against this value without doing central damage to itself, without betraying itself in its innermost self. The absolute demands of the ought correspond to an absolute within the personality, the center of the self (*Wesen*); this implies that it is the center which the absolute demands call on and which receives its essence (*Wesen*) in the decision according to them.
>
> *(Husserl, unpublished manuscripts, in Loidolt, 2012, p. 34)*

For Husserl, the experience of the call arises from within the self, but we might point out that the ethical call does not necessarily come from the center or agency of the self but more primordially from outside of the self. Ethics simply gives itself as a call. In this respect lies the distinction between the moral and the ethical. Moral issues are resolved (through moral theorizing) by arguing from a certain moral perspective the value of the call of the moral predicament or situation that originates within the person. But the ethical transcends the personal.

And yet, a choice may be never wholly chosen:

> Most actual cases are an ambiguous mixture of the chosen and the unchosen, with choice present to a greater or lesser degree. It is necessary for us as agents

to acquire a sense of the nuances and the hints of choice that come and go in what people do: to perceive choice where it may be intimated but is not obvious, to realize that it sometimes may not be present where it seems to be, and to know that such ambiguity occurs in ourselves and in others.

(Sokolowski, 2017, p. 11)

While the ethics of a decision may be revealed in the encounter of having to decide responsibly for an other, we realize the morality of a decision involves senses of good and bad, right and wrong, love and hate, and so forth that ultimately rest with us. However, ethics transcends and, therefore, cannot be reduced to morality. Our seeing, grasping, or otherwise effortful attempts at understanding and acting for this other, whether they be tacit or deliberate, cannot fully attend to the otherness of what is other without reducing otherness to sameness. Can we call any decisions truly ethical?

A difficult decision may haunt the parent and refuse to settle:

I find myself coming back to the decision. I go to brush my teeth, and I find myself thinking about it. It is like a mood that does not seem to go away. Even though I made my decision. I still think of it as not closed, not finished, not done. I still see it as a possibility. This morning we were eating our breakfast, but I just found myself distracted. Their conversation seemed to interrupt me as some kind of deliberation goes on and on. Really, whatever activity I am caught up in—eating, drinking, driving, whatever—I am still in this moment of decision. It is exhausting, but I am just not ready to accept my decision. To remain decided. I am, and I am not, where I need to be. I am back there beside the incubator looking at her. I am back there looking at her face. I am reliving the discussions, reliving the decision. I am revisiting the decision as if unmaking it. It is like the decision is a place that I cannot get settled in but also cannot leave from.

Here is a NICU situation where a parent still seems to live in indecision: distracted, preoccupied, and torn. The decision will not settle as the auras of alternatives remain.

Ethically, we experience the demand to respond to the otherness of others. This is the ethical response in itself; and the becoming of an action as moral rests in how we ultimately realize its goodness, how the act sits with us. This is what Levinas (1981/1974) means when he writes, the "self is goodness" (p. 118). From this perspective, a decision becomes actualized in its morality when it is recognized as subjectively being good or bad. Such understanding does not necessarily occur in the intention or the outcome of the decision but rather whether the decision can unfold to rest with us. A feeling of weight is not just a psychological reaction or emotion but rather expresses our embodied relation with others, the way a moral decision is or is not revealed in its ethicality.

We need to appreciate the weight of moral decisions. Clearly, while there has been a move away from paternalism in decision making, it is recognized that there is a responsibility of health professionals to share in the burden of decision making (Charles et al., 1997). Still, what may be particularly challenging is when we, as healthcare professionals, present decisions as options that in our own eyes do not exist. For example, we may struggle with a family's indecision when we present as a possible decision what we actually perceive as a futile intervention. This is particularly the case when we recognize that conceptually the notion of futility lacks clarity and often wrongly implies certainty. We need to clarify what we deem as futile for it to be meaningfully understood.

The notion of 'futility' has gone in and out of favor over the past decades (Burns & Truog, 2007; Helft et al., 2000). There are of course quantitative dimensions to futility, whereby an intervention may be deemed so unlikely to result in success (i. e. less than 1% chance) such that it is deemed futile. And there are qualitative dimensions, whereby the success that is offered does not meet an acceptable quality of success (i.e. post-intervention quality of life) (Murphy & Finucane, 1993; Schneiderman et al., 1990). Part of the challenge is that individuals may confound medical assessments of futility (*physiological judgment* of futility) with an intervention that offers sufficient benefit to a patient (*value judgment* of futility). For example, mechanical ventilation would not be deemed physiologically futile for the management of respiratory failure yet may be deemed to not offer sufficient benefit to a child with a severe brain injury which has resulted in respiratory failure. While healthcare professionals are poised to evaluate physiological futility, value judgments of futility surely need to be commensurate to nuanced yet also holistic perspectives of a child's best interest.

As a family tries to come to terms with the decision that appears clear to us, we may find ourselves distressed as we continue to provide a treatment that 'sustains life' yet also does not offer a 'cure.' And in the context of providing care often to multiple infants, we may struggle if the resources we devote to so-called 'futile' treatments ultimately compromise the resources available for other children under our care. Other times, we recognize that decisional stress may degenerate into power conflicts whereby health professionals and families are at odds, necessitating an appeal to some higher power (such as the courts) to resolve a decision.

We realize the ethics of moral decisions deals with responsibilities not only to the child but also to the self as an 'other' and others as 'other.' Such an understanding, while illuminating the reality of decision making, also highlights the importance of ongoing research into the ethical moral implications of various healthcare practices. Situations of indecision are likely to become more common as the medical landscape changes, incorporating advanced technologies, often in the context of differing or conflicting cultural values, and so forth. Situations of indecision emphasize the need for a collaborative or even at times delegated decision making, whereby the professional assumes a greater role (Madrigal et al., 2012). Collaborative and delegated decisions may also benefit from policy-level deliberations as resources become limited even in highly technologized healthcare systems.

We must also ultimately acknowledge that ethics expresses living with the weight of fundamental responsibilities towards others that cannot be reduced to hierarchies of values, principles for acting, or other tidy conceptualizations that lend themselves to rationalization or calculation. So, while pragmatically we do need to make moral decisions, the ethos of our professional practice must still comprise a way of dwelling with others whereby the otherness of these others may be encountered in its otherness (Heidegger, 1998/1946).

FALLING INTO DECISIONS

When a child's medical condition continues to get worse and a decision is forced, we may wonder if the decision—uncalculable and unthinkable—may sometimes be resolved not by thinking but rather by falling into it or by taking a sudden leap:

> I do not know at what moment it was, but at some point, I stopped hearing the risks of transplantation, and I finally knew what we had to do. We had been struggling, trying to understand what we should decide. We were going through the process again—how they would qualify her for listing, what being listed would mean, the risks of transplantation, the benefits—all of it. I was listening to what they were saying when I remember suddenly feeling settled. It was done, completed. I was moving on. The decision between palliative care and heart transplantation was made. I had to give her the chance of transplantation no matter the risks, no matter what she may have to go through, even if she died on the waiting list. I had to do that as her mother. I went back to her bedside after the meeting and sat there, looking in her eyes, seeing her face. She was my daughter needing a transplant.

It is at the moment when a decision refuses to settle that suddenly the unexpected may happen. One falls into a decision that suddenly seems utterly decisive and final. We may wonder whether we can remain thoughtful and rational to the end of a decision, or whether in situations that are not resolvable by means of a choice the decision itself is disarticulated to desire, wish, or need. In response to those who view moral decision making as a rational, deliberative affair, we find the reality that often highly consequential decisions are made intuitively (Caeymaex et al., 2001). However, intuitively does not mean automatically, instinctively, or spontaneously. Instead, intuitively speaks to the transcendent unfolding of a decision to the immanent reality of the child and family.

However, the bio-ethical discourse modus continues to emphasize a rational and informed conception of decision making (Batton et al., 2009; Cummings, 2015; Griswold & Fanaroff, 2010; Lemyre et al. 2017; MacDonald, 2002). For example, in informed consent conversations, the emphasis is on: (i) information disclosure, including the nature of the proposed action, the effect of the action, material risks of the action, and the consequences of refusing the action; (ii) capacity to consent, meaning the decision maker's ability to understand the disclosed information; and (iii) voluntariness, a decision maker is able to consent or refuse without duress or coercion. To say that decision making transcends the logic of rationalization does not obfuscate necessary fundamentals. But, a phenomenological account draws attention to the fact that such features do not necessarily define the essence of decision making in its ethics.

We should not be surprised to learn that patient-families vary as far as the amount and type of information they desire (Janvier et al., 2012; Janvier et al., 2014), the preferred processes for deliberation (Caeymaex et al., 2001; Payot et al., 2007), the need for family support (Walter & Ross, 2014; Young et al., 2012), and the accepted presumption of decisional responsibility (Caeymaex et al., 2001; Kon, 2010). To speak of the intuitiveness of decisions does not mean decisions are not multifaceted in their unfolding. Decisions are complex—not reliant on medical information alone or completely rational, at least if we define rationality as in measured or computed risk calculation (Caeymaex et al., 2001; van Manen, 2014b).

The last days have been so hard, ever since we received the diagnosis. The information has been weighing on me. That she would just need so much. And also, that she would have to go through so much. It is not good. It is just not what we want. If we need to let her go, we will. A child should be able to feed. I see her now trying to suck, trying to find something to suck on. Food through a tube is not enough. It is not letting her be a child. Perhaps she is uncomfortable because she knows this is wrong. This is not the life of a child.

Even if decision making may ultimately unfold in intuition (in the sense of direct or immediate cognition), decisions still need deliberation (de Vos et al., 2015; Kon, 2010). As parents come to terms with the reality of their child's current and future situation, the decision unfolds. This unfolding of ethics is not merely the unfolding of the manifolds of complexities composing the decision. It is also the unfolding of the decision itself as the parent comes to terms with what is 'right' for the life of their child. Although contemporary health law may dictate that decisions are made in the 'best interests' of the child, we are aware of the subtle and complex reality of moral decision making that may involve considerations beyond the child: How are the parents to live their own lives? What can they bear? What are the effects of decisions on other important people in their lives, such as their other children and family members?

In neonatal-perinatal medicine, we work in situations where moral consensus may be missing such that at times only perspectives for 'reasonable' courses of action can be offered. Parents may experience decision making in a complex variety of ways (Bogardus et al., 1999; Haward & Janvier, 2015; Janvier et al., 2014; Renjilian et al., 2013; van Manen, 2014b). Insights into parental experiences of decision making need not lead to generalization but instead an acknowledgment that different parents may experience the moments of decisions contingently in different ways. And just as a decision may be a complicated predicament for a parent, much is also demanded from the health professional confronted with the responsibility of helping a family navigate difficult life predicaments (Boss et al., 2009). For some parents, there are decisions that are not quite decisions. And if there is no perceived choice, the healthcare professional's need for a decision may turn into a moment when the parent can only express his or her wants and desires for his or her child. Here, we see the unraveling of the categorical difference between 'what is chosen' and 'what is simply desired.' Although the decision may come from the parent, owned as 'my' decision, we may wonder if sometimes the decision is truly a decision at all, or rather simply a documentation of the wishes of a parent for his or her child. If in the end a parent cannot settle into a decision—necessitating the appeal to technology, nature, or fate—the health professional may find him- or herself drawn deeper into the ethics of the decision.

We need to recognize that what a phenomenological bio-ethics offers to neonatal-perinatal medicine is possible understandings of life meaning. This does not mean that this form of bio-ethics should be relegated to armchair commenting abstractly on the moral issues of medicine. Instead, phenomenological bio-ethics ought to be practiced within the worlds of patients, families, and healthcare professionals to offer questions, reflections, and insights for practice.

CONCLUSION AND BEGINNINGS

The newborn infant does not just enter the pre-existing world of humanity; the newborn is also the creator and the originator of the social and the familial. For neonatal-perinatal medicine, the terminological confluence of 'new' (*neo*) and 'around' (*peri*) to 'birth or origins' (*natalis*) expresses quite precisely the theme of the inceptual beginnings of the child being born (OED). It thematizes how the fetus becomes a child in entering the world. While legally it is only after delivery that the fetus becomes a child, we find ourselves valuing, feeling, assessing, discussing, and making decisions for this not-yet-born before birth. In the practice of neonatal-perinatal medicine, we have to be mindful of the origins and beginnings of the newborn, which prompt us to reflect on the ethics for the newborn, and the origin of this ethics. How can the origins of ethics be traced and understood? While tales of origin and inceptual beginnings can be found in mythology, ancient myths are not just neutral tales from the past. Myths are the bearers of possible worlds, says Paul Ricoeur (1969). They express possible ways of living and of understanding the beginnings and primordial meaning of worlds. Yet, ethics can only be found in myths indirectly.

The Greek poet Hesiod (750–650 BC) tells the geneses, the origin of the world, and the coming into being of the ancient Greek gods (the *Theogony*). From the origin of Chaos, the gaping void, came the first parents: Gaia, the Earth, and Uranus, the Heaven. Their nightly unions bore offspring beginning a line of gods: the three Cyclopes, the three Hecatoncheires, and the twelve Titans. These first children, however, were not simply received by their parents:

> Loathed from the first by their very own father, who, when they were infants,
> Tucked them away in a hole in the earth and prevented their coming
> Up to the light; and Uranus rejoiced in his own evildoing.

Earth, though gigantic, was painfully stuffed on the inside and groaned out Loud.

(Hesiod, 2005, p. 59)

The myth of Gaia and Uranus tells a painful and paradoxical tale. Gaia gave birth to babies who were deformed. Some had a single eye in their forehead, some babies were born with a hundred hands and fifty heads, and other babies possessed unusual power and strength. It was said that Uranus could not accept these offspring. Since Uranus never had a father, he could not possibly know what a father was to do, especially with such unusual children. When a child was born, Gaia could only hold it fleetingly before Uranus seized the newborn to return it back from whence it came in the depths of the earth. The myth is not only beset with deep symbolisms about the origin of humankind; it also confronts us with questions about birth and the taking of responsibility for children. These questions are deeply ethical, but it is puzzling what ethics is supposed to mean in this mythological context as well as for us modern mortals. The important point is that the beginning of birth was also the beginning of ethics. And the beginning of ethics unfolded in the context of the neonatal event: the birth of a child.

As the numerous children were born from the unions of Gaia and Uranus, we can only imagine the ethical predicament of Gaia—her pain, ache, and suffering—as each successive pregnancy carried the expectation and reminder of yet another and all previously failed deliveries. Undoubtedly, Gaia ached from her very depths, where her children were forced to dwell. She could neither see nor touch her newly borns even though their presence was a profound reminder and burden on her body. And so, for Gaia, perhaps the depth of ethical responsivity was not merely the mourning of the loss of each successive newborn infant but also the deeply felt pain of the presence of these children within her earthly embodied being, where each child was doomed to dwell, within the primal womb.

Meanwhile, Uranus remained seemingly untouched by his children and only touched them in a rejecting manner. We might say that he never bore the children in responsivity, neither before nor after birth. While Gaia is commonly regarded as responsible—feeling, caring, attentive—as well as responding to the injustice inflicted on her vulnerable brood, Uranus is seen as hard—cruel, heartless, and callous. Thus, Gaia and Uranus are commonly regarded as symbolizing the presence and absence of primordial ethics. For Gaia, ethics was there from the very beginning: the pain of birthing fused with the ethical pain of losing the newborn child who was doomed to be banished to the deep darkness. But for Uranus, every birth only meant a new threat to his own existence and the exclusivity of his relation to Gaia. In this sense, what is ethical seems to be born out of the responsivity of Gaia, and the lack of responsivity of Uranus.

In view of this grim scenario, the ancient myth of Gaia and Uranus provides an uncanny metaphor for the contemporary importance of neonatal-perinatal ethics. The myth presents a striking image of neonatology: The child must first be received by a parent or other significant caretaker. Indeed, neonatology describes the original birthing encounter of the vulnerable newborn with his or her parent who hopefully takes the child in care—a profoundly ethical moment and perhaps also the moment that calls for moral decisions. But, more specifically, wherein lies the beginning and source of ethics? What sets the beginning of ethics in motion? Can we point at a cause or an event in the process of conception and birth that gave rise to ethics as a human phenomenon? And if so, what does it signify?

Night after night, Uranus coalesced with Gaia. As heaven pressed against earth, there was literally no place on the earthen surface for the children to dwell. From the perspective of Gaia, Uranus was above, the children were below, and she was stuck between. She could not remain present with either. Perhaps, then, it was not without sorrow that Gaia asked her children to do the terrible deed and castrate their father.

> Speaking out boldly, courageously, although afraid in her own heart,
> "My dear children, begotten, alas! of a reprobate father,
> Listen to me and obey: let us punish your father's wrong doing;
> He was the first to conceive of disgraceful and criminal conduct."
> That's what she said. They were all of them frozen by fear, so that no one
> Uttered a word, until mighty, intelligent Cronus took courage,
> And he addressed in the following words his worshipful mother:
> "Mother, I would undertake to accomplish this deed, for I haven't
> Any respect at all for our wretched, unspeakable father,
> Who was the first to conceive such disgraceful and criminal conduct."
>
> *(Hesiod, 2005, p. 59)*

In retelling the myth of Uranus, we may wonder: Can we see more to his action than evil and rejection? Was he so solely filled with hatred and loathing? Or did his burying gesture perhaps oddly reflect some measure of care? After all, Uranus did not destroy his children. He did not commit murder. Instead, he tucked each one away, concealed them, in a safe place: the protective place of their own gestation. Perhaps Uranus' experience, his understanding, only differed from that of Gaia. He could only see things from the perspective of the heavenly firmament.

Uranus, the sky, was hovering above the ethical scene of Gaia and her children. So, he was afforded another perspective, from a heavenly order, to see what was occurring below. Imaginably, it was an observing, examining, or even contemplating look as he appraised the situation, just as he saw it, from a distanced view. To Uranus, the children perhaps appeared unprepared for the world. They were relatively feeble and weak compared to Uranus' own atmospheric immensity and strength. They were really unable to challenge him. Challenge not in the sense of defy; but rather challenge in the sense of *come up* to him as an equal. Gestation had not yet equipped them to thrive.

So, perhaps this primal myth needs a more sensitive hermeneutic re-interpretation that is more understanding of the origin and beginning of ethics. The problem is that the Greek poet Hesiod was known to be misogynistic (Hesiod & Most, 2006), and so his written poem may betray his punishing bias of women. But it is quite possible that, in the depiction of the original oral myths, Uranus experienced the vulnerability of the newborn different from the mother, Gaia. His response to the birth of the child was to put this newborn back where it came from: the womb of mother earth. Subsequent interpreters of this myth have understood this conflict between Gaia and Uranus as the clash between good and evil, compassion and cruelty, kindness and insensitivity. But, perhaps, up to this time in history, we have misjudged Uranus.

Uranus never had a father. And so, he did what an inexperienced father might do. He made a choice of reason. Notwithstanding the suffering that he knew it would cause his partner, Uranus prudently placed immature child after child back into the safety of the womb. Still we may wonder, was this rational choice ethical? And if so, in what sense?

Uranus as the firmament is depicted as a heavenly rational order: above, yet exposed to the down-to-earth encounters below. But his distance from his children is different. He simply cannot descend to feel what is contained within the depths of the earthen world. To come in contact with his children, he would cease being the untouchable sky.

In the mythical encounter of Gaia and Uranus with their newborns, the children do not simply come into being as immaterial objects, without significance, to be accepted or rejected by their parents. The children are not simply bodies, impoverished of meaning, that the mother subjectively intends. Instead, the children have an effect on their parents: effect understood in the fullest sense of sentiment, feeling, and emotion. For in encountering the children, the phenomenality of the phenomenon of neonatology comes forth in the responsivity of the parent as responsibility. And as considered, responsibility as ethics seems to be of a different order for Uranus than for Gaia.

The primordial myth of Gaia and Uranus was meant to let the mortals (human beings) know what gave rise to the beginning of *ethos*. But the contemporary human cannot help but see this as a story that may help explain our human search for meaning, but that still does not account for the beginning of ethics as the beginning of the stirring of life itself. What was there before the beginning? Nowadays, we have a modern, even more powerful narrative, science. And the science of particle physics says, at the beginning there was nothing: Nothing followed by a big bang—the birth of everything! But, of course, this is already a misnomer. Nothingness does not have a before and after. And yet, the birth of ethics is more than an uncanny metaphor that the myth of Uranus and Gaia may represent for us. It is a myth enforced by the cosmological story of science.

Just as the cosmologist speaks of a universe from nothing (Krauss, 2012), so does life seem to start from nothing. We may think that we humans are the cause of this newborn life: after all, as mother and father we feel that we indeed produced this child. But, on second thought, can we really take credit and responsibility for the beginning of this new life? Is that not too presumptuous? Rather, the experiential reality seems to be that we are responsible for a conception and birth that we cannot claim to be responsible for. Ultimately, life was given to us: a self-given gift. Just so with ethics. Ethics gives itself to us as a birth. This means that ethics is not a code or a moral system but rather ethics is the phenomenon that announces itself in a self-giving manner indistinguishable from the conception and birth of the neonatal. Ethics was there even for Uranus, who could not deny it but who did not know what to do with it and tried to put it back again.

From a philosophical point of view, there is a third narrative for understanding the phenomenon and beginning of ethics, and that is the experience of wonder in the face of the other. Ultimately, the experience of life and the ethical as starting from nothingness is just as full of wonder as the alterity of the other that cannot be explained by reducing it to the self or to presence. While cosmology begins with nothingness, phenomenology begins with otherness. Both the empirical science of the nothing and the existential ontology of the other are equally mysterious. The birth of ethics happens in the encounter of the self with the face of the other, says Levinas (1969/1961).

Of course, one might prefer the ethical command of the forbidden fruit of the creation story for the origin of the ethical. Some philosophers draw conclusions about the divinity of conception or about the morality of welcoming or aborting of life, but that moves us into the realm of moral predicaments and moral theories. Here we are still focusing on the ethical, not yet on the moral. Ethics is born in the very instant of the revelation and beginning of this newly conceived life: a beginning without an ostensible beginning and, therefore, a true inceptual event. We have used Heidegger's primordial notion of 'inceptuality' for the beginning and birth of ethics: A true beginning starts from nothing. We can explain the moral with historical and cultural tales and theories that humans have devised, but ethics cannot be explained, except to say that, just like life itself, it can only be explained from its own beginning from nothing, its own self-giving gift in the face of alterity.

The myth of Uranus and Gaia shows that both parents made questionable moral decisions. Uranus made himself the obstacle and violent destroyer of newborn life, infanticide; and Gaia acted to remove the obstacle through the violence of patricide. She charged Cronos to castrate his father by handing him the knife. To make room for the children required a severance of Uranus from Gaia: Cronos had to push him up out of the way, explained Campbell (1949). In the words of Hesiod:

> Taking her son by the hand, Earth hid him in ambush, and put a
> Serrated scythe in his hand, and disclosed to him wholly her dire plot.
> Ushering night in, Uranus visited Gaia, desiring
> Amorous intimacy; he extended himself all around and
> Over the earth, while his son from his ambush protruded his left hand;
> Taking the formidable broad serrated blade in his right, he
> Hastily cut off his own father's privates and cast them behind him.

(Hesiod, 2005, p. 59)

Both Uranus and Gaia, and their son Cronos, became engaged in a fatal series of moral decisions. In finding and fashioning the serrated scythe, the decision and act of revenge for the mishandled life was begun. Uranus, Gaia, and Cronos were all ethically motivated. The problem is that ethics lets us experience (feel and know) that life calls on us, but it does not let us know what to make of the experience. In other words, ethics makes the moral possible, but it cannot tell us what to do.

The fetal beginning of life may only seem latent, not yet touched. But at some point, there is the realization that the pregnant touch is so total, body-within-body that the ethical has come to life and addresses us in its own latent presence. The lesson for healthcare is that we cannot reduce ethics to principles, rules, codes and guidelines of how to live and act morally. But the power and gift of the primordial meaning and significance of the ethical is that it makes it possible to experience ethics as the source of the moral: the ethical requirement for moral reason, decisions, and actions.

The ethical makes it possible to understand the moral. As previously mentioned, the etymology of 'decision' literally means "to cut off" from the Latin *de-caedere* (OED). The decision divides each time, as it is existentially located between what can and cannot be resolved; what is anticipated and unforeseen; and what is tentative and decisive (Derrida, 2008/1999). From the coherence of ethics and decision, it may be surmised that the ethics of a decision expresses more than a cognitive or deliberative affair. A decision cuts between responsivities. So perhaps as Gaia looked down to her confined children, she could not help but look away from Uranus, the sky. The time for deliberation was over, and the decision could not be unmade.

The myth of Gaia and Uranus speaks existentially to the phenomenon of neonatology, the meaning of the ethics of a moral decision. And reflection on this may help raise the issue of the moral decision so that we may gain insight into dimensions of meaning of the ordinary and often extraordinary decisions that parents encounter in the receiving and caring for their children in the neonatal intensive care. We may begin to understand further that the phenomenal meaning of the ethics of a decision lies both in what we may consider as a moral decision and what is ethical in itself. The ethical demand for Gaia is found in an originary responsivity to an other, in the sense of Levinas (1969/1961). Gaia understands not in the mere sense of grasping the meaning of her children's confinement but rather she understands through the existential pain of her motherly suffering.

From this mythical image of the phenomenology of neonatology, we may recognize that ethics emerges from the inceptual encounter of the parent with the child. The meaning of this ethics is not to be found in a formal code of morality that dictates the appropriateness of moral decisions. The ethical arises in the encounter of the parent with the newborn, and from this encounter fundamental decisions are called for. What does it mean to decide? What are the ethics of a decision? What does it mean to make a moral decision?

The legendary myth also shows the newly born children as beings who have been granted life thanks to the technology of the first tool, the serrated sickle, or knife. The children of Gaia and Uranus were vulnerable, in a way premature, unable to survive on their own. It was only with the provision of the technical, the fashioning of the serrated sickle as technology, that the children were made to survive. Yet, to understand the sickle as simply a tool may miss the profound meaning of this act. For with the fashioning of the tool, the children were re-born to emerge from the earthen womb of Gaia but now as technical beings: forever entangled with technology (Stiegler, 1998/1994).

To speak of a technology phenomenologically is to recognize that our being-in-the world is profoundly shaped by technics and technologies. Consider the banal stethoscope.

> As I place the auricle buds in my ears, I find myself oriented to the head of the stethoscope, the diaphragm and the bell. I handle the head ever so carefully to avoid subjecting my ears to the amplified noises that this tool is capable of transmitting. I find myself becoming strangely sensitive to the world as the diaphragm and the bell become my new ears. I am a medical listener. It feels odd to talk or listen to other's voices with the stethoscope on; not only because my voice is altered, but also because it interrupts my focused stethoscope-listening way-of-being. It is as if external voices remind me that I am wearing a device. As I lay the stethoscope against the child's chest, I hear a world of beats, murmurs, and other heart sounds. When I locate the familiar heartbeat, the world available to my other senses seems more distant then the steady rhythm of ta-dum, ta-dum, ta-dum. I am no longer listening to the child so much as I am listening to an organ. The stethoscope changes me and my world.

When a functioning stethoscope is routinely used, we may lose awareness of it as a piece of equipment to only hear the rhythm of the heart. This is not to say that we cannot look at the stethoscope but rather that technologies become unobtrusive in their presence (Heidegger, 1927/1962). Still, even though technologies may become transparent, they do have formative effects on our relation to the world.

In neonatal-perinatal medicine, the child is not necessarily simply given to the parent. The relation is mediated by technologies in various ways (Ihde, 1990). Prior to birth, the gestating child may be seen, heard, and otherwise brought to presence by virtue of technologies. In comparison, in the context of newborn intensive care, it is not just that some of the devices are invasive—tubes physically penetrate bodily surfaces to deliver needed medicines or breathing support—others are non-invasive—wires affixed superficially to measure heart rate, respiratory rate, oxygen saturation, and other vital parameters. Technologies may take the parent relationally closer (or further) from the child. As Albert Borgmann (1984) writes, devices may de-world our relationship with things by disconnecting us from the full actuality of everyday life.

The example of the stethoscope shows that even the simplest useful tool may have ethical implications, as it becomes part of the interpersonal encounter. While it may be entirely necessary for the healthcare professional that the stethoscope presents the child as a 'heart to be heard,' we should wonder about the ethical consequences of neonatal-perinatal technologies that touch on the relation of parent with child. How do various technologies mediate the encounter of parent with child? And how does the application and use of technologies affect what we regard as the ethics of a decision for a child?

> I recall a clinical colleague taking care of a very premature child. The child was extremely fragile; not only in the critical illness sense but also the baby's body itself was fragile—thinly skinned, softly boned, and delicately featured. As the child became sicker, it became necessary to view the heart to gain a sense of its function. My colleague appropriately obtained an ultrasound machine, unbundled the child to expose the chest, and pressed the probe against the sternum to look at the filling, contractility, and ejection fraction of the organ. When she was satisfied, having finished the ultrasound study, she removed the probe. Then she looked at the chest: it was bruised. The pressure of the probe had been too much. She was horrified by what she had done.

How is it that the professional touches and is touched by the child? On the one hand, the professional needs to probe to understand and manipulate pathophysiologic processes. And on the other hand, he or she must also stand back to relate to the actual patient he or she is responsible for.

The image of Cronos provides a metaphor for the modern human as a technical being. In the NICU, neither the health professional nor the parent's ethical experiences of the child can be fully understood without also considering the various effects that the different medical technologies may infuse and latently carry into the encounter between self and other. For example, the child may come forth shaped by the visibility of the ultrasound probe, a modern sickle, not as a child but instead as a physiological heart to be seen. And it may only be in laying down the probe that the child may be experienced in unmediated responsibility. Again, while it may be tempting to consider technologies as simply either present or absent, we might recall how even the most non-invasive technologies, such as the neonatal cardiorespiratory monitor, influence the parent's sensibilities in complex ways (van Manen, 2015).

So, while the focus of this text has been on exploring the ethical sense of the child from the perspectival experience of the parent, we should also remember that these technologies are formative of the experience of the professional and likely profoundly formative of the experiences of the newborn child in ways that we do not yet fully recognize and understand. Still, we should not look at technology despairingly. The modern human is not only constituted by the technological; humans may also (re)design the technologies that constitute them: These are ethical-moral acts.

As technical human beings, professionals, or parents, we are afforded to use, design, or built technologies that not only affect us and our relation to the other but also the other and his or her relations to yet others. In this sense, we may be responsible for the latent effects of technologies on others. We may wonder how perinatal technologies affect not only the professional and the parent but also the becoming of the child. What opportunities do they promote or discourage? What are their existential significances for the child of the past, the now and the child of the future? What are the consequences of beginning life in this contemporary, technological world?

And so we see that while Gaia's decision was ultimately born from responsivity, in the end it became an ethically animated moral act. These profound ethical-moral questions need further consideration and study in various clinical contexts. A critical thoughtfulness towards technology is clearly required whereby we do not only view technologies instrumentally but also as formative of our very humanity. And while the answers are unlikely to be solely founded within phenomenological inquiry, phenomenology may help to identify, explicate, and raise these issues. Phenomenology may reawaken us to the primordial ethics that is born con-temporaneously with the new life.

In the various ways that technologies may mediate our relation to the world, they shape our sensibilities—creating a pattern, an atmosphere, an environment of perception that works on us. And by awakening our sensibilities, the technologies of neonatal-perinatal medicine may change the pace and interaction between parent and child.

ETHICAL MORAL PERSPECTIVES

Medical ethics and bio-ethics has turned into a complex field with respect to theoretical perspectives, rational moral principles, discipline methodologies, the variety of ethical issues, as well as policy matters and political consequences. As well, the scope of the field of medical ethics and bio-ethics is not agreed upon, just as how the terms medical ethics and bio-ethics express distinct subject matters.

The concept of ethics is used quite broadly and variously (Cahn & Markie, 1998). Distinctions can be made between meta-ethics, which deals with the question of where ethical principles come from and what they mean; normative ethics, which asks how we ought to conduct ourselves in practice; and applied ethics, which examines specific controversial issues (Fieser, 2000). The divisions, however, are not marked by clear boundaries with each area presupposing understandings of the others. In this text, the term 'bio-ethics' is used not only for the field concerned with classical medical ethical questions revolving around the provision of medical treatments and use of medical technologies, but also recognizing that this field, in a fundamental way, is concerned with the *ethos* of medical practice.

In neonatal-perinatal medicine, a manifold of ethical moral situations can arise—for example, regarding consent for surgical treatment, confidentiality of medical information, or truth telling following events of medical error. Ethics reside in high-stake situations such as decision making for infants born at the cusp of viability, withdrawing medical interventions for infants with severe brain injury, or experimental therapies for infants with uncertain prognoses. Ethics are also present in the daily practices of clinicians and their interactions with patients and their families. And yet, sometimes, for understandable and appropriate reasons, people are primarily preoccupied with the bio-technical aspects of clinical practice such that they pass over ethics. Alternatively, individuals can become so fixated with seeing a situation from a particular perspective of ethics that alternative interpretations are hard to discern.

There is value in considering some of the existing moral philosophies and ethical theories routinely employed in neonatal-perinatal medicine. On the one hand, we can regard these differing ethical moral perspectives as complementary to each other, offering varied understandings of ethical moral situations. Their value is not necessarily founded in the possibility of reaching resolution of ethical moral predicaments but rather in offering differing perspectives and insights. On the other hand, we can also view each perspective as effecting a reduction of the meaning of ethics and potentially offsetting other perspectives. Clearly, a bio-ethics that becomes overly abstract or focuses exclusively on theoretical cases risks becoming a so-called 'quandary ethics' whereby the ethics of concrete, real-world situations are passed over, just as perspectives that seek to universalize cases to principles risk abstracting that which is precisely ethical—falling into moral solipsism.

> Moral solipsism is self-refuting. If the self is wrested from the intersubjective context of concrete moral action, then it becomes an abstracted and 'lifeless' self which is neither moral nor immoral. It becomes a formal determinant in a logical or epistemological scheme of things and is divested of its existential reality. The moral consciousness in its concretion is indelibly communal.
>
> *(Schrag, 2013/1963, p. 93)*

When a clinician reduces moral actions to a deductive rationalization of duties or harms, the ethical relatedness of patient, family, and clinicians partaking in decision making may quite simply fail to be seen. This is concerning if we believe that the ethics of neonatal-perinatal medicine are ultimately founded in the face-to-face encounters between healthcare practitioners, families, and the infants.

Juxtaposing the various ethical moral traditions shows how none on their own ultimately (in)forms a comprehensive grasp of ethics and morality. The language of each, their way of seeing a situation, cannot help but introduce certain constraints or perspectives that may make it difficult to discuss the ethics of neonatal-perinatal medicine in nuanced yet necessary terms. Still, each ethical moral perspective is potentially complementary to each other.

Deontological Perspectives

A deontological perspective, sometimes described as obligation-, duty-, or rule-based ethics, views an action as right or wrong based on how it conforms to what binds it. 'Deontology,' from the Greek *deon*, expresses obligation as "that which is binding, duty" (OED). There are numerous formulations of deontological ethics, with moral obligations arising from varied sources. For example, central to Judaism, Christianity, and Islam are the Decalogue, the Ten Commandments, and the Sayings respectively, each consisting of prohibitions and prescriptions such as 'thou shalt not kill' and 'honor thy father and mother.'

Immanuel Kant argued that to act in a morally right way people must act from duty. It is not the consequences of actions that confer morality to actions but the motives of the people who carry out the actions. Good consequences, after all, could foreseeably arise from malevolent motivations or even chance. Kant's famous first formulation of the 'categorical imperative' expresses an absolute and unconditional requirement for action based on reason: "Act only according to that maxim whereby you can, at the same time, will that it should become a universal law" (Kant, 1994/1785, p. 30). For Kant, actions are realized as 'right' based on how they confirm to this reason such that any rational person would and, therefore, should treat imperatives as universal specifying duties. In neonatal–perinatal medicine, clinicians should act in such a way that they think they would want others to act if they were providing care to them.

In contemporary bio-ethics, the four *Principles of Biomedical Ethics* are designed to function as the cornerstones for a duties-based deontological ethics (Beauchamp & Childress, 2001). These principles are commonly articulated and put to work to analyze so-called ethical moral problems. In short, a moral action should do good (beneficence), avoid harm (non-maleficence), respect the independence of the patient (autonomy), and be equitable not just for the patient, or group of patients, but also for society at large (justice). While healthcare professionals have long been obligated to promote welfare and avoid harm, the challenge in neonatal medicine is that it can be difficult to discern what interests an infant has or will have—let alone agree on what is truly good and what is of unacceptable harm. More so, infants are also without autonomy, relying on their parents, based on their special relationship, to make decisions for them. And yet, autonomy means more than freedom to make decisions as surrogate decision makers. As Meyers (2004) wrote:

> genuine autonomy entails more than the mere making of decisions; it requires both the capacity to make free and informed decisions and the active development of character by which persons understand and are able to act upon self-defining choices ... autonomy undercutting power asymmetries prevail and decision making in routine care relies much more on *assent* than on *consent* ... healthcare in general, and critical care in particular, represent profoundly difficult context for genuinely autonomous choices.
>
> *(p. 105)*

The final principle of justice tends to be invoked in situations wherein health professionals are obligated to treat recipients of an action equitably. Clinicians should not discriminate based on age, gender, social standing, and so forth. This principle is also related to social justice, which refers to fair and equitable distribution of resources. These are of course not the only principles that can be articulated to guide practice; other principles may be invoked for crafting healthcare guidelines and procedures. For example, resource allocation policies frequently appeal to the principles of utility, equity, and solidarity for rationing healthcare resources in situations of scarcity.

A different side to principles is rights: One person's rights are another person's obligations. At times, the notion of rights has been given greater emphasis than duties, particularly in political spheres. Consider the United Nations Universal Declaration of Human Rights (1948), or the Convention on the Rights of the Child (United Nations General Assembly, 1989). Rights are sometimes easier to voice than to uphold. A rights-based deontological ethics becomes quickly problematic when we realize that individual rights may conflict with one another, or even the rights of others.

Criticisms to deontology-based approaches express frustration that it is precisely in ethically challenging situations that deontology seems to fall short. How do we act when we are unclear or disagree upon what is beneficial for a particular patient? What harms are too much for an infant to bear? Can we liken respecting the autonomy of a family to the autonomy of the newborn? Alternatively, it can be argued that identified obligations, duties, or rules are often used superficially, as slogans or rationale, to support convictions that simply conform to an individual's deeply held beliefs.

Although obligations, duties, or rules make it possible to formulate consistent justifications, it can be argued that there is ultimately no real ground for any obligation, duty, or rule. Deontology-based approaches ultimate rest on an assumption of 'moral fundamentalism,' meaning that it is possible and appropriate to articulate 'basic' or 'fundamental' moral principles across contexts and cultures. Such a position collapses under cultural and contemporary social science critiques. Finally, deontology runs the risk of not necessarily being benevolent given that morality from a deontological perspective is indifferent to the consequences of actions.

Consequentialist Perspectives

Consequentialism takes the consequences of an act as the ultimate basis for moral judgment about the rightness or wrongness of that conduct. It is sometimes termed teleological ethics, from the Greek *telos*, meaning "the end, limit, goal, fulfillment, completion" (OED). Judgments about actions may be made prospectively, in terms of how well designed they are to produce good outcomes, or, retrospectively, according to whether or not they actually produced such outcomes. Like deontology, there are varied forms of consequentialisms. Dating back to at least 5th century BC, the Mohists believed that morality is based on "promoting the benefit of all under heaven and eliminating harm to all under heaven" (Fraser, 2011, p. 62). Such an articulation of consequentialism places moral weight on what is good for the community rather than individualistic goals.

Utilitarianism names a family of consequentialist ethical theories that endorses actions that maximize utility, which is often defined in terms of happiness, well-being, or some related concept. For a hedonistic utilitarianism, the moral action is the one judged to produce the greatest happiness for the greatest number, while the worst is the one that causes the most misery. Classically, the interests of all are considered equal, unlike in other versions of consequentialism such as egoism and altruism.

Consequentialism is a natural fit for many clinicians, as clinical decision making involves weighing advantages and risks, gains and harms, benefits and burdens. It is in line with the medical literature that clarifies, elaborates, and appraises decisions based on rational decision making theory. Although consequentialism is often placed in opposition to deontological perspectives, consequentialist theories can encompass or appeal to rules and principles such as rule-based utilitarianism, which we find in the balancing of beneficence and non-maleficence.

In neonatal-perinatal medicine, an appeal is made to the so-called 'best interests of the child,' which can be used to define utility. What constitutes 'best interests' is more holistic than a medical determination of interests, so that terms such as well-being are often appealed to. Still, wellbeing is not without ambiguity, consideration, and debate when it comes to conceptions of quality of life or other outcomes. How much brain injury constitutes such harm that continuing treatment would be immoral? How do interests of the infant balance against those of the family and society when weighing options? How do we deal with uncertainty in outcomes? Health practitioners and families may disagree on what constitutes the 'best' best interests, so an appeal may need to be made to elucidate actions that yield 'reasonable' (rather than necessarily 'best') outcomes. Finally, a deeply held criticism of consequentialist perspectives is expressed by the idiomatic phrase, 'the ends do not justify the means.' We do not need to look back very far in humanity's history to find abhorrent actions performed for particular ends.

Character Perspectives

Person-based approaches to ethics, such as virtue ethics, shift the focus of attention from the action to the agent. Morality is judged based on the virtuous qualities of the individual: disposition and character. The notion of virtue can be traced to the Greek word for excellence, *arête*. In its basic sense, *arête* means "excellence" of any kind; literally, "that which is good" (OED). Yet for the Greeks, this notion of excellence was bound up with the notion of the fulfillment of purpose or function: the act of living up to one's full potential.

We recognize virtues in the characters of story, myth, and legend. Traditional western virtues are courage, self-control, justice, wisdom, faith, hope, and love; vices are pride, avarice, lust, envy, gluttony, wrath, and sloth. The kind of person we are goes hand in hand with the actions we pursue, to the extent that a virtuous person acts virtuously in their nature, their actions fulfill their purpose. Although virtue ethics orients us to the person, we actually realize virtues in their enactment. The courageous behave courageously; the just acts justly; the wise displays wisdom, and so forth. How else would we realize their virtue?

In neonatal-perinatal medicine, a virtue ethics perspective leads us to ask what qualities make a good physician, nurse, social worker, or other health professional. And of the parent of the child, what qualities make a good parent? We can also ask comparative questions such as are the virtues of a physician different from a nurse as far as fulfilling their professional purpose? The point about virtues is that they are not reducible to obligations, principles, or other deontological rules that guide actions. Rather, virtues are the qualities that characterize an individual and are bound up in their roles as healthcare professionals or parents. Virtues give rise to behaviors, habits, routines, or other actions unfolding in the context of clinical practical practice.

Virtues are indications of the ethical character of a person. We recognize the qualities of a good nurse, a good doctor, and so forth as those features that trainees in such professions ought to emulate and teachers ought to cultivate. Clearly, in team-based medical practice, virtuous clinicians may be recognized and sought out particularly in situations that are morally or ethically challenging.

Challenges to virtue approaches include that questions remain unanswered: Who constitutes a good role model? Do certain virtues supersede other virtues? How do we consider opposing actions of so-called virtuous individuals? We also recognize that in some situations virtues may be vices and the contrary—vices may promote good actions.

Communitarian Perspectives

Communitarian derives from the Old French *comuner*, "to make common, share" in *comun* "common, general, free, open, public" (OED). A narrative communitarian ethics focuses on understanding morality through story and particular events in the lives of individuals and communities. At the core of narrative ethics lies the presumption that moral truths are founded and expressed in narratives. A communitarian narrative is not simply a story encompassing characters and events but a means of sense-making of human life, and specifically moral life. In neonatal-perinatal medicine, we can gain an understanding of ethics by paying attention to how a narrator is situated within a story; a story's beginning, plot, and ending and the relationships between what is said and what is heard in the telling of a story. We seem able to accept those moral situations which we can talk about as 'good' stories, not in the sense of entertaining stories but rather moral stories that we can recount without a sense of regret, despair, or a sense of wrong-doing.

Relational ethics takes as a starting point that ethical practice takes place within the context of human relationships, and as such, systematic analysis of ethical issues needs to consider these relationships (Bergum & Dossetor, 2005). At its core, relational ethics asserts that human flourishing is enhanced by healthy and ethically sound relationships. For neonatal-perinatal medicine, the parent is viewed in relation not only with his or her child but also in the social world of friends, family, and healthcare professionals. A proper understanding of parental relational experiences may foster ethical sensitivity in the practitioner. Relational ethics may also expand the discussion or understanding of ethical issues and frameworks as attention is extended beyond the patient as the focus of ethical moral questions.

Ethics of care was originally described by Gilligan (1982) and subsequently developed by various feminist theorists. This approach foreshadows the interdependency of givers and receivers of care in ethical moral situations as appreciated by relational ethics. It also considers those virtues expressed in caring relation-ships—compassion, interdependence, respect, and responsibility—as center-points for exploring moral understandings.

Contractualist Perspectives

Contractualism posits that we create morality by our agreements with one another. From this perspective, prior to such agreements or shared understandings, actions are neither inherently moral nor immoral. Instead, what is constituted as moral is recognized to be under provisional resolution as agreed-upon rules are formed and re-formed. In a contract, we are drawn together, from *com* "with, together" and *trahere* "to draw" (OED).

Contractualist perspectives are associated with the social and political theory of Thomas Hobbes. Informed consent can be seen as a contractualist ethics where moral actions are judged in accord with mutual agreements of the decision makers. Consent may be obtained before a child is born for situations where it is unclear whether medical interventions will be in a child's interest (i.e. situations of extreme prematurity, multiple congenital anomalies, and so forth). And consent may be obtained before blood product transfusions, invasive procedures, and, of course, surgery. An action is judged morally right to the extent that it aligns with agree-ments made between healthcare professionals and patients or patients' legal guar-dians. In neonatal-perinatal medicine, parents are generally charged with authority as substitutive decision makers for their children.

Much of intensive care, however, is pursued without explicit expression and documentation of consent. Instead, the contract to care is implicit in its practice. And yet, contractualist perspectives of ethics hinge on whether a genuine contract was formed. Is truly informed consent possible when it comes to counseling for medical interventions? Do families really appreciate what they are consenting to when they pursue resuscitation of an infant at the threshold of viability? Do they understand what intensive care will look like, the possible short-term risks, and the possible long-term consequences? It may also happen that at times 'the state' is asked by health professionals to intervene, to act *in loco parentis*, when a family's decision seems to lie outside of reasonableness, or when a family otherwise lacks capacity to give consent.

Casuistry and Pragmatic Perspectives

Casuistry is an example of a perspective-based approach whereby morality is not viewed as something 'out there' but rather residing in the particularities of a 'case.' The moral question is, about the 'case' in its essential form, the Latin *cadere*, "to fall" (OED). What is the case? And what has befallen?

Casuistry dates from Aristotle. It reflects a process of reasoning that seeks to resolve moral problems by extracting and extending theoretical rules from a parti-cular case and reapplying those rules to new instances. A moral problem is, therefore, viewed as an interpretation of events shaped by history, tradition, and precedent. The case at hand is considered against the backdrop of a history of similar and dissimilar cases. Casuistry is the method of jurisprudence. Related to casuistry is pragmatic ethics, which treats morality much like a science, such that moral criteria are subjected to reinterpretation and revision over time.

In neonatal-perinatal medicine, it is certainly not uncommon to appeal to previous cases as a means of asserting that because in similar cases a certain approach was viewed as appropriate that same approach in the case at hand is reasonable to pursue. Alternatively, there are also those negative or problematic cases that haunt, disturb, or otherwise shape future practices as individuals reflect on what happened and whether it should be allowed to happen again. Critics of casuistry and pragmatic ethics see bio-ethics as moral relativism or bio-ethics unraveling into the past shaping the future, meaning a possible naïve acceptance that because something was accepted to be moral it is morally right to continue to act in the same way moving forward.

Phenomenological Perspectives

A phenomenological ethics, as articulated in this book, reflects on concrete situations to arrive at an understanding of ethics as expressed in the manner we directly experience the world—rather than in terms of some abstract principles or consequences of an action.

> To look at ethical involvement from the neutral standpoint of theoretical reasoning is not the right phenomenological way to approach this issue. Rather, the ethical experience itself has to be described in a way that it captures the lived experience of being called, of being addressed by an "ought" itself. In doing this, phenomenological description exceeds the dichotomy between "the descriptive" and "the normative."
>
> *(Loidolt, 2012, p. 14)*

Every healthcare situation has its own particular hospital staff, illness situation, patient and family, and other nuanced complexities.

A phenomenological ethics starts from the concrete yet also reflects beyond particulars to explore the phenomenological meaning of phenomena. For example, in the context of neonatal-perinatal medicine, multiple technologies are routinely used: mechanical ventilators and other respiratory devices to support oxygen delivery and carbon dioxide clearance; tube enteral and parenteral nutrition to provide nutrition for growth; and so forth. While deontological and consequential perspectives look at such interventions as 'means' affording actions (judged as moral based on their underlying principles or consequences, respectively), a phenomenological ethics draws attention to the meaning of living through such interventions. Breathing is not reducible to respiration understood physiologically as oxygen delivery or carbon dioxide removal; just as eating is not reducible to provision of enteral or parenteral nutrition. A phenomenological ethics, therefore, draws us to consider the experiential life we are affording critically ill babies and the lives of families and hospital staff. Such an ethics is particularly important when we recognize that for some infants their hospital stay is more than days or even weeks; some babies dwell in intensive care for many months only to eventually depart the hospital dependent on medical technologies.

A phenomenological ethics cannot help but abut against language, as a phenomenologist realizes that when we speak of an action, a person, or even a situation as being 'good' or 'bad,' we cannot reduce moral judgments to rational argument. For example, following Wittgenstein, we can speak of a good ventilator as one that aptly oxygenates a child, but the morality of a breathing machine is more than its utility.

On a good ventilator, a child does not appear to struggle against the machine. There is no asynchrony of machine pushing a breath in while the child tries to exhale. Instead, his or her chest takes on a restful cadence of inhalation and exhalation such that hopefully the machine becomes unseen in its presence. More so, what is good is expressed in our understanding of the machine as relieving gasping, panting, or other signs of breathlessness. We hope for the child that breathing is in itself effortless and unnoticed. We are perhaps, then, more morally comfortable with a child who lives on breathing support when we have a sense that this other life, this life of the child, is one without suffering or one where suffering has been relieved.

And yet, a phenomenological ethics needs to recognize that this other life, the life of the child, may differ in its subjectivity from our own. So while we cannot approach the life of an other from a perspective other than our own, we may still wonder about the possible meaningfulness of a new life starting dependent on medical technology (van Manen, 2019). While the breathing machine may support breathing to be effortless, the breathing tube acts as a barrier, blocking the child's mouth. We know that newborns encounter their world through their mouths: their mother's breast, their own hands, and other objects. When a breathing tube obstructs the mouth, what is the newborn's new oral sensuality? How does he or she come to know his or her world? While a phenomenological ethics cannot necessarily answer these questions, it can raise our awareness to their ethical meaning and significance.

REFERENCES

Alderson, P., Hawthorne, J., & Killen, M. (2006). "Parents' experiences of sharing neonatal information and decisions: consent, cost and risk." *Social Science & Medicine*, 62(6), 1319–1329.

Alldred, S. K., Takwoingi, Y., Guo, B., Pennant, M., Deeks, J. J., Neilson, J. P., & Alfirevic, Z. (2017). "First trimester ultrasound tests alone or in combination with first trimester serum tests for Down's syndrome screening." *Cochrane Database of Systematic Reviews*, 3, CD012600.

Avery, M. E. (1992). "A 50-year overview of perinatal medicine." *Early Human Development*, 29(1–3), 43–50.

Barry, M. J., & Edgman-Levitan, S. (2012). "Shared decision-making: the pinnacle of patient centered care." *New England Journal of Medicine*, 366(9), 780–781.

Batton, D. G.*et al.* (American Academy of Pediatrics, Committee on Fetus and Newborn). (2009). "Clinical report—antenatal counseling regarding resuscitation at an extremely low gestational age." *Pediatrics*, 124(1), 422–427.

Beard, M. (2018, February 3). "Is having a baby unethical?" *ABC News*. Retrieved from https://www.abc.net.au/news/2018-02-04/philosophy-parenting-is-having-a-baby-u nethical/9382564.

Beauchamp, T. L., & Childress, J. F. (2001). *Principles of Biomedical Ethics* (5th ed.). New York, NY: Oxford University Press.

Beaudry, J. S. (2016). "Beyond (models of) disability?" *Journal of Medicine and Philosophy*, 41(2), 210–228.

Bergum, V. (1997). *A Child on Her Mind: The Experience of Becoming a Mother*. Westport, CT: Bergin & Garvey.

Bergum, V., & Dossetor, J. B. (2005). *Relational Ethics: The Full Meaning of Respect*. Hagerstown, MD: University Publishing Group.

Blanco, F., Suresh, G., Howard, D., & Soll, R. F. (2005). "Ensuring accurate knowledge of prematurity outcomes for prenatal counseling." *Pediatrics*, 115(4), e478–487.

Bogardus, S., Holmboe, E., & Jekel, J. (1999). "Perils, pitfalls and possibilities in talking about medical risk." *JAMA*, 281(11), 1037–1041.

Borgmann, A. (1984). *Technology and the Character of Contemporary Life: A Philosophical Inquiry*. Chicago, IL: University of Chicago Press.

Boss, R. D., Hutton, N., Donohue, P. K., & Arnold, R. M. (2009). "Neonatologist training to guide family decision making for critically ill infants." *Archives of Pediatrics & Adolescent Medicine*, 163(9), 783–788.

Boss, R. D., Hutton, N., Sulpar, L. J., West, A. M., & Donohue, P. K. (2008). "Values parents apply to decision-making regarding delivery room resuscitation for high-risk newborns." *Pediatrics*, 122(3), 583–589.

Boström, E. J. (2016). "The unborn child and the father: acknowledgement and the creation of the other." In J. Bornemark & N. Smith (Eds.), *Phenomenology of Pregnancy*. Södertörn Philosophical Studies (pp. 141–156). Stockholm, SE: Elanders.

Brock, D. W. (1996). "What is the moral authority of family members to act as surrogates for incompetent patients?" *Milbank Quarterly*, 74(4), 599–618.

Buchanan, A. E., & Brock, D. W. (1989). *Deciding for Others: The Ethics of Surrogate Decision Making*. Cambridge, UK: Cambridge University Press.

Burns, J. P., & Truog, R. D. (2007). "Futility: a concept in evolution." *Chest*, 132(6), 1987–1993.

Buytendijk, F. J. J. (1970). "Some aspects of touch." *Journal of Phenomenological Psychology*, 1(1), 99–124.

Caeymaex, L., Speranza, M., Vasilescu, C., Danan, C., Bourrat, M. M., Garel, M., & Jousselme, C. (2001). "Living with a crucial decision: a qualitative study of parental narratives three years after the loss of their newborn in the NICU." *PLoS One*, 6(12), e28633.

Cahn, S. M., & Markie, P. (Eds.). (1998). *Ethics: History, Theory, and Contemporary Issues*. New York, NY: Oxford University Press.

Callahan, D. (2003). "Individual good and common good: a communitarian approach to bioethics." *Perspectives in Biology and Medicine*, 46(4), 496–507.

Campbell, J. (1949). *The Hero with a Thousand Faces*. Princeton, NJ: Princeton University Press.

Cavalo, A., Dierckx de Casterlé, B., Naulaers, G., & Gastmans, C. (2020). "Ethics of resuscitation for extremely premature infants: a systematic review of argument-based literature." *Journal of Medical Ethics*. E-pub ahead of print. doi:10.1136/medethics-2020-106102.

Charles, C., Gafni, A., & Whelan, T. (1997). "Shared decision-making in the medical encounter: what does it mean? (or it takes at least two to tango)." *Social Science & Medicine*, 44(5), 681–692.

Chervenak, F. A., Skupski, D. W., Romero, R., Myers, M. K., Smith-Levitin, M., Rosenwaks, Z., & Thaler, H. T. (1998). "How accurate is fetal biometry in the assessment of fetal age?" *American Journal of Obstetrics & Gynecology*, 178(4), 678–687.

Committee on Hospital Care. (2003). "Family centered care and the pediatrician's role." *Pediatrics*, 112(3 Pt 1), 691–696.

Cummings, J.*et al.* (American Academy of Pediatrics, Committee on Fetus and Newborn). (2015). "Antenatal counseling regarding resuscitation and intensive care before 25 weeks of gestation." *Pediatrics*, 136(3), 588–595.

Davis, L., Mohay, H., & Edwards, H. (2003). "Mothers' involvement in caring for their premature infants: an historical overview." *Journal of Advanced Nursing*, 42(6), 578–586.

De Beauvoir, S. (1976/1948). *The Ethics of Ambiguity*. New York, NY: Kensington Publishing.

De Beauvoir, S. (2004/1945). "A review of the phenomenology of perception." (M. Timmermann, Trans.). In M. A. Simons (Ed.), *Simone de Beauvoir: Philosophical Writings* (pp. 151–164). Urbana, IL: University of Illinois Press.

De Vos, M. A., Bos, A. P., Plötz, F. B., van Heerde, M., de Graaff, B. M., Tates, K., Truog, R. D., Willems, D. L. (2015). "Talking with parents about end-of-life decisions for their children." *Pediatrics*, 135(2), e465–476.

De Vries, R. (2017). "Obstetric ethics and the invisible mother." *Narrative Inquiry in Bioethics*, 7(3), 215–220.

Deber, R. B. (1994). "Physicians in health care management: 8. The patient-physician partnership: decision-making, problem solving and the desire to participate." *Canadian Medical Association Journal*, 151(4), 423–427.

Derrida, J. (1976/1967). *Of Grammatology*. (G. C. Spivak, Trans.). Baltimore, MD: Johns Hopkins University Press.

Derrida, J. (2000/2005). *On Touching—Jean-Luc Nancy*. (C. Irizarry, Trans.). Stanford, CA: Stanford University Press.

Derrida, J. (2008/2009). *The Gift of Death and Literature in Secret*. (D. Wills, Trans.). Chicago, IL: University of Chicago Press.

Diekema, D. S. (2011). "Revisiting the best interest standard: uses and misuses." *Journal of Clinical Ethics*, 22(2), 128–133.

Dondorp, W. J., Page-Christiaens, G. C., & De Wert, G. M. (2016). "Genomic futures of prenatal screening: ethical reflection." *Clinical Genetics*, 89(5), 531–538.

Donohoe, J. (2012). "The phenomenological shift of parenthood." In M. Sanders & J. J. Wisnewski (Eds.), *Ethics and Phenomenology* (pp. 185–196). Plymouth, UK: Lexington Books.

Doron, M. W., Veness-Meehan, K. A., Margolis, L. H., Holoman, E. M., & Stiles, A. D. (1998). "Delivery room resuscitation decisions for extremely premature infants." *Pediatrics*, 102(3), 574–582.

Duff, R. S., & Campbell, A. G. M. (1973). "Moral and ethical dilemmas in the special-care nursery." *New England Journal of Medicine*, 289(17), 890–894.

Dunstan, G. R. (1984). "The moral status of the human embryo: a tradition recalled." *Journal of Medical Ethics*, 10(1), 38–44.

Dupont-Thibodeau, A., Barrington, K. J., Farlow, B., & Janvier, A. (2014). "End of life decisions for extremely low-gestation-age infants: why simple rules for complicated decisions should be avoided." *Seminars in Perinatology*, 38(1), 31–37.

Dworkin, R. (1994). *Life's Dominion: An Argument About Abortion, Euthanasia, and Individual Freedom*. New York, NY: Vintage Books.

Ferrarello, S. (2016). *Husserl's Ethics and Practical Intentionality*. London, UK: Bloomsbury Academic.

Fieser, J. (2000). *Metaethics, Normative Ethics, and Applied Ethics: Contemporary and Historical Readings*. Belmont, CA: Wadsworth, Cengage Learning.

Finnis, J. (1994). "Abortion and health care ethics." In R.E. Ashcroft*et al.*, *Principles of Health Care Ethics* (pp. 547–557). Chichester, UK: John Wiley.

Fox, R. C. (1990). "The evolution of American bioethics: a sociological perspective." In G. Weisz (Ed.), *Social Science Perspectives on Medical Ethics* (pp. 201–217). Dordrecht, NL: Kluwer Academic Publishers.

Fraser, C. (2011). "Mohism and legalism." In J. L. Garfield & W. Edelglass, *The Oxford Handbook of World Philosophy* (pp. 58–67). Oxford, UK: Oxford University Press.

Gaucher, N., & Payot, A. (2011). "From powerlessness to empowerment: mothers expect more than information from the prenatal consultation for preterm labour." *Paediatrics & Child Health*, 16(10), 638–642.

Gaucher, N., Nadeau, S., Barbier, A., Janvier, A., & Payot, A. (2016). "Personalized antenatal consultations for preterm labor: responding to mothers' expectations." *Journal of Perinatology*, 178, 130–134.e7.

Gilligan, C. (1982). *In a Different Voice.* Cambridge, MA: Harvard University Press.

Gilmore, A. (1984). "Sanctity of life verses quality of life: the continuing debate." *Canadian Medical Association Journal,* 130(2), 180–181.

Giorgi, A. (1970). *Psychology as a Human Science: A Phenomenologically Based Approach.* New York, NY: Harper & Row.

Giorgi, A. (2009). *The Descriptive Phenomenological Method in Psychology: A Modified Husserlian Approach.* Pittsburgh, PA: Duquesne University Press.

Godolphin, W. (2003). "The role of risk communication in shared decision-making." *British Medical Journal,* 327(7417), 692–693.

Griswold, K. J., & Fanaroff, J. M. (2010). "An evidence-based overview of prenatal consultation with a focus on infants born at the limits of viability." *Pediatrics,* 125(4), e931–937.

Habermas, J. (2003/2001). *The Future of Human Nature.* (W. Rehg, M. Pensky, & H. Beister, Trans.). Cambridge, UK: Polity Press.

Hall, S. L., Hynan, H. T., Phillips, R., Lassen, S., Craig, J. W., Goyer, E., Hatfield, R. F., & Cohen, H. (2017). "The neonatal intensive parenting unit: an introduction." *Journal of Perinatology,* 37(12), 1259–1264.

Harrison, H. (1993). "The principles for family-centered neonatal care." *Pediatrics,* 92(5), 643–650.

Haward, M. F., & Janvier, A. (2015). "An introduction to behavioural decision-making theories for paediatricians." *Acta Paediatrica,* 104(4), 340–345.

Haward, M. F., Gaucher, N., Payot, A., Robson, K., & Janvier, A. (2017). "Personalized decision making: practical recommendations for antenatal counseling for fragile neonates." *Clinics in Perinatology,* 44(2), 429–445.

Haward, M. F., Murphy, R. O., & Lorenz, J. M. (2008). "Message framing and perinatal decisions." *Pediatrics,* 122(1), 109–118.

Heidegger, M. (1927/1962). *Being and Time.* (J. Macquarrie & E. Robinson, Trans.). New York, NY: Harper & Row.

Heidegger, M. (1971). *Poetry, Language, Thought.* New York, NY: Harper & Row.

Heidegger, M. (1977). *The Question Concerning Technology and Other Essays.* New York, NY: Harper & Row.

Heidegger, M. (1998/1946). "Letter on 'humanism'." (F. A. Capuzzi, Trans.). In W. McNeill (Ed.), *Pathmarks* (pp. 239–276). Cambridge, UK: Cambridge University Press.

Helft, P. R., Siegler, M., & Lantos, J. (2000). "The rise and fall of the futility movement." *New England Journal of Medicine,* 343(21), 293–296.

Hesiod. (2005). *Works of Hesiod and the Homeric Hymns.* (D. Hine, Trans.). Chicago, IL: University of Chicago Press.

Hesiod, & Most, G. W. (2006). *Hesiod.* Cambridge, MA: Harvard University Press.

Hume, D. (1738). *A Treatise of Human Nature.* London, UK: John Noon.

Husserl, E. (2005). *Phantasy, Image Consciousness and Memory (1898–1925).* Dordrecht, IL: Springer.

Ihde, D. (1990). *Technology and the Lifeworld: From Garden to Earth.* Bloomington, IN: Indiana University Press.

Ihde, D. (2002). *Bodies in Technology.* Minneapolis, MN: University of Minnesota Press.

Introna, L. D. (2019). "Ethics and the speaking of things." *Theory, Culture & Society,* 26(4), 25–46.

Janvier, A., & Farlow, B. (2015). "The ethics of neonatal research: an ethicist's and a parents' perspective." *Seminars in Fetal and Neonatal Medicine,* 20(6), 436–441.

Janvier, A., & Lantos, J. (2016). "Delivery room practices for extremely preterm infants: the harms of the gestational age label." *Archives of Disease in Childhood. Fetal and Neonatal Edition,* 101(5), F375–376.

Janvier, A., Baardsnes, J., Hebert, M., Newell, S., & Marlow, N. (2017). "Variation of practice and poor outcomes for extremely low gestation births: ordained before birth." *Archives of Disease in Childhood. Fetal and Neonatal Edition*, 102(6), F470–471.

Janvier, A., Barrington, K., & Farlow, B. (2014). "Communication with parents concerning withholding or withdrawing of life sustaining interventions in neonatology." *Seminars in Perinatology*, 38(1), 38–46.

Janvier, A., Farlow, B., Baardsnes, J., Pearce, R., & Barrington, K. J. (2016). "Measuring and communicating meaningful outcomes in neonatology: a family perspective." *Seminars in Perinatology*, 40(8), 571–577.

Janvier, A., Lorenz, J. M., & Lantos, J. D. (2012). "Antenatal counselling for parents facing an extremely preterm birth: limitations of the medical evidence." *Acta Paediatrica*, 101(8), 800–804.

Jennings, B. (1990). "Ethics and ethnography." In G. Weisz (Ed.), *Social Science Perspectives on Medical Ethics* (pp. 261–272). Dordrecht, NL: Kluwer Academic Publishers.

Jonas, H. (1984). *The Imperative of Responsibility: In Search for an Ethics for the Technological Age.* Chicago, IL: Chicago University Press.

Jonsen, A. J. (1980). "Can an ethicist be a consultant?" In V. Abernathy (Ed.), *Frontiers in Medical Ethics* (pp. 157–171). Cambridge, MA: Ballinger.

Jonsen, A. J. (1986). "Casuistry and clinical ethics." *Theoretical Medicine*, 7(1), 65–73.

Kant, I. (1994/1785). *Ethical Philosophy: The Complete Texts of Grounding for the Metaphysics of Morals and Metaphysical Principles of Virtue (Part II of The Metaphysics of Morals) with On a Supposed Right to Lie Because of Philanthropic Concerns* (2nd ed.). (J. W. Ellington, Trans.). Indianapolis, IN: Hackett Publishing Company.

Kavanaugh, K., Savage, T., Kilpatrick, S., Kimura, R., & Hershberger, P. (2005). "Life support decisions for extremely premature infants: report of a pilot study." *Journal of Pediatric Nursing*, 20(5), 347–359.

Klein, E. (1971). *A Comprehensive Etymological Dictionary of the English Language.* New York, NY: Elsevier Scientific Publishing Company.

Kon, A. A. (2010). "The shared decision-making continuum." *JAMA*, 304(8), 903–904.

Kopelman, A. E. (2006). "Understanding, avoiding, and resolving end-of-life conflicts in the NICU." *Mount Sinai Journal of Medicine*, 73(3), 580–586.

Kopelman, L. M. (2007). "The best interests standard for incompetent or incapacitated persons of all ages." *The Journal of Law, Medicine & Ethics*, 35(1), 187–196.

Krauss, L. M. (2012). *A Universe from Nothing: Why There Is Something Rather than Nothing.* New York, NY: Free Press.

Kushnir, V. A., Barad, D. H., Albertini, D. F., Darmon, S. K., & Gleicher, N. (2017). "Systematic review of worldwide trends in assisted reproductive technology 2004–2013." *Reproductive Biology and Endocrinology*, 15(1), 6.

Ladd, R. E., & Mercurio, M. R. (2003). "Deciding for neonates: whose authority, whose interests?" *Seminars in Perinatology*, 27(6), 488–494.

Lago, P. M., Piva, J., Garcia, P. C., Troster, E., Bousso, A., Sarno, M. O., Torreão, L., Sapolnik, R. (Brazilian Pediatric Center of Studies on Ethics). (2008). "End-of-life practices in seven Brazilian pediatric intensive care units." *Pediatric Critical Care Medicine*, 9(1), 26–31.

Landzelius, K. (2003). "Humanizing the imposter: object relations and illness equations in the neonatal intensive care unit." *Culture Medicine and Psychiatry*, 27(1), 1–28.

Lantos, J. D. (2001). *The Lazarus Case: Life-and-Death Issues in Neonatal Intensive Care.* Baltimore, ML: John Hopkins University Press.

Lantos, J. D. (2018). "What is the purpose of antenatal counseling?" *Journal of Pediatrics*, 196, 8–10.

Lemyre, B.*et al.* (Canadian Paediatric Society, Fetus and Newborn Committee). (2017). "Counselling and management for anticipated extremely preterm birth." *Paediatrics & Child Health*, 22(6), 334–341.

Lerman, C., Daly, M., Walsh, W. P., Resch, N., Seay, J., Barsevick, A., Birenbaum, L., Heggan, T., & Martin, G. (1993). "Communication between patients with breast cancer and health care providers: determinants and implications." *Cancer*, 72(9), 2612–2620.

Leuthner, S. R. (2001). "Decisions regarding resuscitation of the extremely premature infant and models of best interest." *Journal of Perinatology*, 21(3), 193–198.

Levinas, E. (1969/1961). *Totality and Infinity: An Essay on Exteriority.* (A. Lingis, Trans.). Pittsburgh, PA: Duquesne University Press.

Levinas, E. (1981/1974). *Otherwise than Being or Beyond Essence.* (A. Lingis, Trans.). The Hague, NL: Duquesne University Press.

Levinas, E. (1989/1949). "The transcendence of words." In S. Hand (Ed.), *The Levinas Reader* (pp. 144–149). Oxford, UK: Blackwell Publishers.

Liberman, R. F., Getz, K. D., Heinke, D., Luke, B., Stern, J. E., Declercq, E. R., Chen, X., Lin, A. E., & Anderka, M. (2017). "Assisted reproductive technology and birth defects: effects of subfertility and multiple births." *Birth Defects Research*, 109(14), 1144–1153.

Lingis, A. (1994). *Foreign Bodies.* New York, NY: Routledge.

Løgstrup, K. E. (1997/1956). *The Ethical Demand.* Notre Dame, IN: University of Notre Dame Press.

Loidolt, S. (2012). "The 'Daimon' that speaks through love: a phenomenological ethics of the absolute ought. Investigating Husserl's unpublished ethical writings." In M. Sanders & J. J. Wisnewski (Eds.), *Ethics and Phenomenology* (pp. 9–38). Plymouth, UK: Lexington Books.

MacDonald, H.*et al.* (American Academy of Pediatrics, Committee on Fetus and Newborn). (2002). "Perinatal care at the threshold of viability." *Pediatrics*, 110(5), 1024–1027.

Madrigal, V. N., Carroll, K. W., Hexem, K. R., Faerber, J. A., Morrison, W. E., & Feudtner, C. (2012). "Parental decision-making preferences in the pediatric intensive care unit." *Critical Care Medicine*, 40(10), 2876–2882.

Malenka, D. J., Baron, J. A., Johansen, S., Wahrenberger, J. W., & Ross, J. M. (1993). "The framing effect of relative and absolute risk." *Journal of General Internal Medicine*, 8(10), 543–548.

Marion, J.-L. (2000). "The voice without name: homage to Levinas." In J. Bloechl (Ed.), *The Face of the Other and the Trace of God: Essays on the Philosophy of Emmanuel Levinas* (pp. 224–242). New York, NY: Fordham University Press.

Marron, J. M., Jones, E., & Wolfe, J. (2018). "Is there ever a role for the unilateral do not attempt resuscitation order in pediatric care?" *Journal of Pain and Symptom Management*, 55 (1), 164–171.

Martin, J. A., Hamilton, B. E., Osterman, M. J. K., Driscoll, A. K., & Drake, P. (2018). "Births: final data for 2017." *National Vital Statistics Reports*, 67(8), 1–50.

Mazur, D. J., & Hickman, D. H. (1991). "Patients' interpretations of probability terms." *Journal of General Internal Medicine*, 6(3), 237–240.

McHaffie, H. E. (2001). *Crucial Decisions at the Beginning of Life: Parents' Experiences of Treatment Withdrawal from Infants.* Abingdon, UK: Radcliffe Medical Press.

Merleau-Ponty, M. (1962/1945). *Phenomenology of Perception.* (C. Smith, Trans.). New York, NY: Routledge & Kegan Paul.

Merleau-Ponty, M. (1968/1964). *The Visible and the Invisible.* (A. Lingis, Trans.). Evanston, IL: Northwestern University Press.

Merleau-Ponty, M. (1993). "Eye and mind." In G. A. Johnson (Ed.), *The Merleau-Ponty Aesthetics Reader: Philosophy and Painting* (pp. 121–162). Evanston, IL: Northwestern University Press.

Mesman, J. (2008). *Uncertainty and Medical Innovation in Neonatal Care: Experienced Pioneers in Neonatal Care*. Basingstoke, UK: Palgrave MacMillan.

Meyers, C. (2004). "Cruel choices: autonomy and critical care decision-making." *Bioethics*, 18(2), 104–119.

Milan, A. (2014). "Fertility: fewer children, older moms." *Statistics Canada*. Catalogue no. 11–630-X.

Miller, G. (2006). *Extreme Prematurity: Practices, Bioethics, and the Law*. Cambridge, UK: Cambridge University Press.

Milunsky, A., & Milunsky, J. (Eds.). (2016). *Genetic Disorders and the Fetus: Diagnosis Prevention, and Treatment*. Hoboken, NJ: John Wiley.

Miquel-Verges, F., Woods, S. L., Aucott, S. W., Boss, R. D., Sulpar, L. J., & Donohue, P. K. (2009). "Prenatal consultation with a neonatologist for congenital anomalies: parental perceptions." *Pediatrics*, 124(4), e573–579.

Mirmiran, M., Barnes, P. D., Keller, K., Constantinou, J. C., Fleisher, B. E., Hintz, S. R., & Ariagno, R. L. (2004). "Neonatal brain magnetic resonance imaging before discharge is better than serial cranial ultrasound in predicting cerebral palsy in very low birth weight preterm infants." *Pediatrics*, 111(4), 992–998.

Morain, S., Greene, M. F., & Mello, M. M. (2013). "A new era in noninvasive prenatal testing." *New England Journal of Medicine*, 369(6), 499–501.

Mumford, J. (2013). *Ethics at the Beginning of Life: A Phenomenological Critique*. Oxford, UK: Oxford University Press.

Murphy, D. J., & Finucane, T. E. (1993). "New do-not-resuscitate policies: a first step in cost control." *Archives of Internal Medicine American Medical Association*, 153(14), 1641–1648.

Nelson, J. L. (1992). "Taking families seriously." *The Hastings Center Report*, 4(22), 6–12.

Norton, M. E. (2013). "Follow-up of sonographically detected soft markers for fetal aneuploidy." *Seminars in Perinatology*, 37(5), 365–369.

OED. *Online Etymology Dictionary*. www.etymonline.com.

Parens, E. (2009). "Respecting children with disabilities—and their parents." *The Hastings Center Report*, 39(1), 22–23.

Payot, A., Gendron, S., Lefebvre, F., & Doucet, H. (2007). "Deciding to resuscitate extremely premature babies: how do parents and neonatologists engage in the decision?" *Social Science & Medicine*, 64(7), 1487–1500.

Perlman, N. B., Freedman, J. L., Abramovitch, R., Whyte, H., Kirpalani, H., & Perlman, M. (1991). "Informational needs of parents of sick neonates." *Pediatrics*, 88(3), 512–518.

Philip, A. G. S. (2005). "The evolution of neonatology." *Pediatric Research*, 50(4), 799–815.

Pinch, W. J. E. (2002). *When the Bough Breaks: Parental Perceptions of Ethical Decision-Making in NICU*. Lanham, MD: University Press of America.

Prescott, S., & Hehman, M. C. (2017). "Premature infant care in the early 20th century." *Journal of Obstetric, Gynecologic, & Neonatal Nursing*, 46(4), 637–646.

President's Commission. (1983). *Deciding to Forego Life-Sustaining Treatment: A Report on the Ethical, Medical and Legal Issues in Treatment Decisions/President's Commission for the Study of Ethical Problems in Medicine and Biomedical and Behavioral Research*. Washington, DC: The Commission.

Pugliese, A. (2016). "Phenomenology of drives: between biological and personal life." In J. Bornemark & N. Smith (Eds.), *Phenomenology of Pregnancy*. Södertörn Philosophical Studies (pp. 71–90). Stockholm, SE: Elanders.

Raffoul, F. (2010). *The Origins of Responsibility*. Bloomington, IN: Indiana University Press.

Renjilian, C. B., Womer, J. W., Carroll, K. W., Kang, T. I., & Feudtner, C. (2013). "Parental explicit heuristics in decision-making for children with life-threatening illnesses." *Pediatrics*, 131(2), e566–572.

Rhodes, R., & Holzman, I. R. (2014). "Is the best interest standard good for pediatrics?" *Pediatrics*, 134(Suppl 2), S121–129.

Ricoeur, P. (1966/1950). *Freedom and Nature: The Voluntary and the Involuntary*. (E. V. Kohák, Trans.). Evanston, IL: Northwestern University Press.

Ricoeur, P. (1969). *The Symbolism of Evil*. New York, NY: Beacon Press.

Ricoeur, P. (1977/1975). *The Rule of Metaphor*. (R. Czerny, K. McLaughlin, & J. Costello, Trans.). Toronto, CA: University of Toronto Press.

Rieder, T. N. (2017). "Saving or creating: which are we doing when we resuscitate extremely preterm infants?" *American Journal of Bioethics*, 17(8), 4–12.

Roelants-van Rijn, A. M., Groenendaal, F., Beek, F. J., Eken, P., Van Haastert, I. C., & De Vries, L. S. (2001). "Parenchymal brain injury in the preterm infant: comparison of cranial ultrasound, MRI and neurodevelopmental outcome." *Neuropediatrics*, 32(2), 80–89.

Roscigno, C. I., Savage, T. A., Kavanaugh, K., Moro, T. T., Kilpatrick, S. J., Strassner, H. T., Grobman, W. A., & Kimura, R. E. (2012). "Divergent views of hope influencing communications between parents and hospital providers." *Qualitative Health Research*, 22 (9), 1232–1246.

Ryan, C. A., Byrne, P., Kuhn, S., & Tyebkhan, J. (1993). "No resuscitation and withdrawal of therapy in a neonatal and a pediatric intensive care unit in Canada." *Journal of Pediatrics*, 123(4), 534–538.

Saltvedt, S., Almström, H., Kublickas, M., Reilly, M., Valentin, L., & Grunewald, C. (2004). "Ultrasound dating at 12–14 or 15–20 weeks of gestation? A prospective cross-validation of established dating formulae in a population of in-vitro fertilized pregnancies randomized to early or late dating scan." *Ultrasound in Obstetrics & Gynecology*, 24(1), 42–50.

Sands, R., Manning, J. C., Vyas, H., & Rashid, A. (2009). "Characteristics of deaths in paediatric intensive care: a 10-year study." *Nursing in Critical Care*, 14(5), 235–240.

Sartre, J.-P. (2004/1940). *The Imaginary: A Phenomenological Psychology of the Imagination*. (J. Webber, Trans.). New York/London: Routledge.

Schneiderman, L. J., Jecker, N. S., & Jonsen, A. R. (1990). "Medical futility: its meaning and ethical implications." *Annals of Internal Medicine*, 112(12), 949–954.

Schrag, C. O. (2013/1963). *Reflections on the Religious, the Ethical, and the Political*. Lanham, MD: Lexington Books.

Seo, B. K. (2016). "Caring for premature life and death: the relational dynamics of detachment in a NICU." *Medical Anthropology*, 35(6), 560–571.

Simms, E. M. (2008). *The Child in the World: Embodiment, Time, and Language in Early Childhood*. Detroit, MI: Wayne State University Press.

Slovic, P. (Ed.). (2001). *The Feeling of Risk*. London, UK: Earthscan.

Sokolowski, R. (2017). *Moral Action: A Phenomenological Study*. Washington, DC: The Catholic University of America Press.

Spivak, G. C. (1995). "Translator's preface." In M. Devi, *Imaginary Maps*. (G. C. Spivak, Trans.) (pp. xxiii–xxix). New York, NY: Routledge.

Stiegler, B. (1998/1994). *Technics and Time, 1: The Fault of Epimetheus*. Stanford, CA: Stanford University Press.

Studer, E. M., & Marc-Aurele, K. L. (2016). "Lost in explanation: lessons learned from audio-recordings and surveys of the antenatal consultation." *Journal of Neonatal-Perinatal Medicine*, 9(4), 393–400.

Sunderam, S., Kissin, D. M., Zhang, Y., Folger, S. G., Boulet, S. L., Warner, L., Callaghan, W. M., & Barfield, W. D. (2019). "Assisted reproductive technology surveillance—United States, 2016." *Morbidity and Mortality Weekly Report–Surveillance Summaries*, 68(4), 1–23.

Svenaeus, F. (2017). *Phenomenological Bioethics: Medical Technologies, Human Suffering, and the Meaning of Being Alive*. Abingdon, UK: Routledge.

Tooley, M. (1972). "Abortion and infanticide." *Philosophy and Public Affairs*, 2(1), 37–65.

Tørring, N. (2016). "First trimester combined screening – focus on early biochemistry." *Scandinavian Journal of Clinical and Laboratory Investigation*, 76(6), 435–447.

Tyson, J. E., Parikh, N. A., Langer, J., Green, C., & Higgins, R. D. (National Institute of Child Health and Human Development Neonatal Research Network). (2008). "Intensive care for extreme prematurity: moving beyond gestational age." *New England Journal of Medicine*, 358(16), 1672–1681.

United Nations General Assembly. (1948). *Universal Declaration of Human Rights*. United Nations, 217 A (III).

United Nations General Assembly. (1989). *Convention on the Rights of the Child*. United Nations, Treaty Series, 1577, p. 3.

Van Manen, M. (1990). *Researching Lived Experience: Human Science for an Action Sensitive Pedagogy*. London, ON: Althouse Press.

Van Manen M. (1999). "The pathic nature of inquiry and nursing." In I. Madjar & J. Walton (Eds.), *Nursing and the Experience of Illness: Phenomenology in Practice* (pp. 17–35). London, UK: Routledge.

Van Manen, M. (2014a). *Phenomenology of Practice: Meaning-Giving Methods in Phenomenological Research and Writing*. New York, NY: Routledge.

Van Manen, M. (2002). "Care-as-worry, or 'don't worry, be happy'." *Qualitative Health Research*, 12(2), 262–278.

Van Manen, M. A. (2012a). "Technics of touch in the neonatal intensive care." *Medical Humanities*, 38(2), 91–96.

Van Manen, M. A. (2012b). "Looking into the neonatal isolette." *Medical Humanities*, 38(1), e4.

Van Manen, M. A. (2012c). "Ethical responsivity and pediatric parental pedagogy." *Phenomenology & Practice*, 6(1), 5–17.

Van Manen, M. A. (2014b). "On ethical (in)decisions experienced by parents of infants in neonatal intensive care." *Qualitative Health Research*, 24(2), 279–287.

Van Manen, M. A. (2015). "The ethics of an ordinary medical technology." *Qualitative Health Research*, 25(7), 996–1004.

Van Manen, M. A. (2019). *Phenomenology of the Newborn: Life from Womb to World*. New York, NY: Routledge.

Verbeek, P. P. (2008). "Obstetric ultrasound and the technological mediation of morality: a postphenomenological analysis." *Human Studies*, 31(1), 11–26.

Verbeek, P. P. (2011). *Moralizing Technology: Understanding and Designing the Morality of Things*. Chicago, IL: University of Chicago Press.

Verhagen, A. A., Janvier, A., Leuthner, S. R., Andrews, B., Lagatta, J., Bos, A. F., & Meadow, W. (2010). "Categorizing neonatal deaths: a cross-cultural study in the United States, Canada, and The Netherlands." *Journal of Pediatrics*, 156(1), 33–37.

Verkerk, M. A., & Lindemann, H. (2014). "End-of-life decisions for newborns." In J. D. Arras, E. Fenton, & R. Kukla (Eds.), *The Routledge Companion to Bioethics* (pp. 500–510). New York, NY: Routledge.

Vollmer, B., Roth, S., Baudin, J., Stewart, A. L., Neville, B. G., & Wyatt, J. S. (2003). "Predictors of long-term outcome in very preterm infants: gestational age versus neonatal cranial ultrasound." *Pediatrics*, 112(5), 1108–1114.

Von Hauff, P., Long, K., Taylor, B., & van Manen, M. A. (2016). "Antenatal consultation for parents whose child may require admission to neonatal intensive care: a focus group study for media design." *BMC Pregnancy Childbirth*, 16, 103.

Waldenfels, B. (2010). "Responsive ethik zwischen antwort und verantwortung." [Responsive ethics between response and responsibility]. *Deutsche Zeitschrift fur Philosophie*, 58(1), 71–81.

Waldenfels, B. (2011/2006). *Phenomenology of the Alien: Basic Concepts*. (T. Stähler & A. Kozin, Trans.). Evanston, IL: Northwestern University Press.

Walter, J., & Ross, L. F. (2014). "Relational autonomy: moving beyond the limits of isolated individualism." *Pediatrics*, 133 (Sup 1), S16–23.

Weiss, P., & Taruskin, R. (Eds.). (1994). *Music in the Western World: A History in Documents*. Belmont, CA: Thomson Schirmer.

Wilkinson, D. (2013). *Death or Disability? The Carmentis Machine and Treatment Decisions for Critically Ill Children*. Oxford, UK: OUP.

Wilkinson, D., De Crespigny, L., & Xafis, V. (2014). "Ethical language and decision-making for prenatally diagnosed lethal malformations." *Seminars in Fetal and Neonatal Medicine*, 19 (5), 306–311.

Yee, W. H., & Sauve, R. (2007). "What information do parents want from the antenatal consultation?" *Paediatrics & Child Health*, 12(3), 191–196.

Young, E., Tsai, E., & O'Riordan, A. (2012). "A qualitative study of predelivery counseling for extreme prematurity." *Paediatrics & Child Health*, 17(8), 432–436.

Young, I. M. (1984). "Pregnant embodiment: subjectivity and alienation." *The Journal of Medicine and Philosophy: A Forum for Bioethics and Philosophy of Medicine*, 9(1), 45–62.

INDEX

abortion 23, 37
antenatal consultation 31, 39, 42–43
attachment 60–62, 67–70
autonomy 19, 52, 75, 77, 89, 110–111

best interest 2, 31, 43, 75, 86, 89–91, 94, 97, 112

casuistry 75–76, 83, 114–115
chance 17–18, 22, 35, 37–44, 77, 80–81, 94, 110; *see also* decision making; *see also* risk
communitarian 84, 113–114
conception 3, 7, 15–20, 23, 35, 42, 51, 101–103
congenital anomalies 9, 19, 27, 30
consent 81, 88, 97, 110, 114
consequentialism 43, 83–84, 111–112, 115
contractualism 114

decision making: conflict 4, 84, 94, 102; conversational 43; decisional responsibility 97–98; end-of-life 79; improvised 43; message framing 41; personalized 43; shared 75, 94; uncertainty 25–26, 76, 82, 85, 87, 90
delivery room 5, 31
deontology 83–84, 109–111, 115
disability 36, 39–44, 83

embodiment 42
emotions 4, 9, 28, 30, 50, 88, 93, 102
ethics of care 113–114

expectations 5, 18, 28–29, 33–38, 88–89; *see also* hope

family-centred care 76
futility 79, 94

genetic condition and testing 35–37
guidelines 104, 110

hope 2, 5, 18, 28, 33–38, 41, 50, 64–65, 76, 81, 85

imaging technologies: computed tomography 63; magnetic resonance imaging 63–66, 80; ultrasound 27–32, 35–36, 40, 63, 77, 82, 85, 106
infertility 19
innovation 9, 19, 58

justice 7, 32, 100, 110

labor 1, 6, 37, 73–74

moral distress 80, 94

narrative ethics 113–114

paternalism 75, 94
patient-centred care 76
personhood 11, 26, 37, 74
phenomenology 9–10, 12, 28, 103, 107
pragmatism 114–115
pregnancy 1, 6–7, 13–44, 51, 100, 104,

prematurity 18–19, 41, 51–52, 54–55, 58, 62, 65, 68, 105–106
prenatal testing 17, 36–37
principlism 42, 74, 104, 108, 110–111

quality of life 15, 37, 39–44, 75, 79, 82–83, 94, 112

relational ethics 113–114
religion and spiritual 4–5, 78
reproductive technologies 19–20
responsibility 2, 7, 11, 19, 25, 32, 36, 48–50, 56–57, 67–70, 76, 78, 84, 90–92, 94, 100, 102, 106
risk 2, 6, 18–19, 35–38, 40–41, 58, 80, 83, 85, 96–97, 112 , 114

sanctity of life 78–79

surrogate decision maker 75, 110

technology 2, 5, 7, 28–32, 36–38, 48, 51, 53–58, 61–66, 94, 98, 104–107, 115–116

uncertainty 25, 41, 76, 82, 85, 87, 90, 112
utilitarianism 111–112
utility 58, 110, 111–112, 116

values 10, 40, 76, 80, 84, 88–89, 94–95
viability and threshold of viability 2, 6, 27, 37, 39, 58
virtue and virtue ethics 4, 80, 83, 112–113
vitalism 78–79

withdrawing of medical interventions 77–78, 80, 82–84
withholding of medical interventions 52

For Product Safety Concerns and Information please contact our EU
representative GPSR@taylorandfrancis.com
Taylor & Francis Verlag GmbH, Kaufingerstraße 24, 80331 München, Germany

www.ingramcontent.com/pod-product-compliance
Lightning Source LLC
Chambersburg PA
CBHW070735220326
41598CB00024BA/3434